AF764706

Water Resources Planning, Development and Management

Editor

Keith Marcell

scitus academics

Water Resources Planning, Development and Management
Edited by **Keith Marcell**

ISBN: 978-1-68117-619-2
Library of Congress Control Number: 2016946477

© 2017 by
SCITUS Academics LLC,
616, Corporate Way, Suite 2, 4766,
Valley Cottage, NY 10989

www.scitusacademics.com

This book contains information obtained from highly regarded resources. Copyright for individual articles remains with the authors as indicated. All chapters are distributed under the terms of the Creative Commons Attribution License, which permits unrestricted use, distribution, and reproduction in any medium, provided the original author and source are credited.

Notice

Reasonable efforts have been made to publish reliable data and views articulated in the chapters are those of the individual contributors, and not necessarily those of the editors or publishers. Editors or publishers are not responsible for the accuracy of the information in the published chapters or consequences of their use. The publisher believes no responsibility for any damage or grievance to the persons or property arising out of the use of any materials, instructions, methods or thoughts in the book. The editors and the publisher have attempted to trace the copyright holders of all material reproduced in this publication and apologize to copyright holders if permission has not been obtained. If any copyright holder has not been acknowledged, please write to us so we may rectify.

Printed in United States of America on Acid Free Paper

Preface

Water, at the forefront of global policy change, planning and management, is a gradually important issue. Water as a renewable resource and its availability to a wide range of users, aquatic ecosystem health and global issues related to climate change, water security, water trade and water ethics are increasingly concerned. The management of water resources is to plan, develop, manage and distribute the maximum use of water resources. It is a sub-set of management of water cycles. Ideally, the planning of water resources management takes into account each of the competing water requirements and seeks to allocate water on an equitable basis to meet all requirements and uses. In practice, this is rarely possible, as with other resource management.

Water Resources Planning, Development and Management is a compendium of innovative up-to-date perspectives on key aspects of water resources planning, development and management that are important to professionals and researchers alike. Successful management of any resources requires accurate information on the available resources, the uses to which they could be put, competing resource requirements, measures and processes to assess the importance and value of competing mechanisms and requirements to translate policy decisions into action on the ground. Much work in the management of water resources aims to optimize water use and minimize the environmental impact of water use on the natural environment.

Table of Contents

Chapter 1

Recent Developments in Monitoring of Microbiological Indicators of Water Quality Across a Range of Water Types

Sandra Mesquita[1] and Rachel T. Noble[2]

[1] Center Centre for Marine Sciences from Algarve (CCMAR), Campus de Gambelas, Portugal

[2] Institute of Marine Sciences, University of Chapel Hill, North Carolina, Morehead City, USA

1. INTRODUCTION

As human pressure increases, so does exploitation of natural water sources for production of drinking water, public use and appreciation of recreational waters. Subsequently, the demand for faster and reliable monitoring methods and approaches has also been intensified. Ensuring that water is safe, whether for bathing or consumption is a critical complex process that requires the participation of multiple stakeholders [1, 2] and also distinct degrees of knowledge with respect to water management.

For quantifying microbial contaminants that are dangerous to public health, fecal indicator bacteria (FIB) are used worldwide as indicators of the potential presence of dangerous pathogens that can be found in water used for bathing, drinking and harvesting of seafood. The “indicator” system relies on the assumption that FIB are present concomitant to the presence of human bacterial, viral and parasitic pathogens of concern. Furthermore, the indicator system simplifies management, because no single pathogen can be singularly attributed to the majority of waterborne disease. If pathogens were measured for management of water quality, the list of candidates to be quantified would be many. The FIB system relies on thresholds that have been established using epidemiological studies and risk assessment frameworks (see Annapolis Protocol and WHO bathing water guidelines).While the standards for drinking water, recreational water and shellfish harvesting waters and meats are based upon similar approaches [3], different types of waters can have different acceptable thresholds for the same FIB because of differences in either acceptable risk, the relative use of the water (i.e. for consumption versus recreation), and a range of other factors. Therefore, acceptable thresholds for FIB in drinking water, recreational water and shellfish harvesting waters and meats vary by continent, region, state and locality.

By definition an indicator (in this case, FIB) should be, 1) able to be enumerated using standardized tests that are relatively cost-effective and user-friendly, 2) present in high concentrations in human fecal contamination, matching the high concentrations of other respective pathogens, and 3) well studied in relation to human health, with epidemiological studies to link FIB concentrations and adverse health outcomes such as gastrointestinal and respiratory human diseases [4, 5, 6]. These assumptions can be tenuous at times for all types of waters and for all of the FIB groups currently utilized for water quality management. However, it is generally accepted that regulation using FIB to protect public health is successful. For example, total coliforms do not exactly fit these requirements since they are not exclusively fecal in origin; however they are often used in conjunction with other FIB enumeration approaches to protect public health in drinking water [7]. Even though some of the assumptions listed above are violated periodically for the use of FIB to protect public health in the regulation of drinking water, an alternative approach is often unattainable. Human pathogens are heterogeneous, can be highly seasonal, and can be highly infectious at low doses. The range of enteric viruses alone that could be monitored in drinking water is vast, and these pathogens are difficult to detect [8, 9, 11]. A discussion of the pathogens of concern is beyond the scope of this chapter, and the reader is referred to a range of quality publications on the topic (e.g. 11, 12).The "indicator" system can be used successfully, particularly if new research and methodologies in the water quality arena are incorporated into management. As a community, we can attain improved means of management of public health in water, if attention is paid to utilization and incorporation of new monitoring tools, new models and molecular tools. Current advances in science include rapid quantitative PCR assays (QPCR) developed to track distinct microorganisms in multiple water sources [12, 13, 14, 15, 16, 7].

In this chapter, we highlight recent advancements in the management of water, and in monitoring of microbial water quality in drinking water production and recreational waters. First, we highlight the current regulatory landscape for microbiological indicators. Second, we will describe the application and evolution of molecular tools for rapid quantification of FIB (E. coli and *Enterococcus* sp.), in a range of water types. We will include some of the limitations and advantages of different types of culture based methods as compared to the recently developed rapid QPCR-based methods. We will present a small case study conducted along high use beaches along the Pacific Ocean of California, USA. Third, we highlight recent developments in the field of Harmful Algal Blooms (HABs) and concerns over cyanobacterial toxins in surface water sources for drinking water. Following the presentation of information presented on HABs, and comparing *E. coli* and *Enterococcus*sp. concentrations using culture based and molecular (qPCR) rapid methods, the authors will consider the application of newly developed source tracking markers for quantification of fecal contamination fecal contamination. These microbial source tracking (MST) tools are currently being used and expanded for the management of a range of water types. The chapter is intended to highlight recent advancements, while promoting the consideration of other successful approaches for managing water resources well into the 21st century with an eye toward improved protection of public health.

2. REGULATORY LANDSCAPE ACROSS MAJOR WATER TYPES: HIGHLIGHTING THE USE OF THE FECAL INDICATOR BACTERIA AS PART OF THE "INDICATOR" SYSTEM

Traditionally, the application of FIB for water quality monitoring included enumeration of members of the total coliform group. This group was originally described on the basis of lactose fermentation detection instead of upon the principles of systematic bacteriology [3; 17-26]. Total coliforms are still used widely for management of drinking water, but their sole use in recreational water quality monitoring has decreased in favor of enumeration of FIB that are specific to fecal contamination from warm-blooded animals (i.e. *E. coli* and *Enterococcus* sp.). Total coliforms are defined by their respective characteristics, i.e. they are Gram negative facultative anaerobes that do not form spores are rod-shaped bacteria, with lactose fermentation occurring with acid production (24-48h, 36°C), and are indole-negative. Coliforms belong to the family Enterobacteriaceae which includes *Escherichia, Enterobacter, Klebsiella* and *Citrobacter, Kluyvera, Leclercia*, and some members of the genus*Serratia* [27].

In all of the culture-dependent methods, cultivation conditions are choose in order to improve the growth of the target microorganism while simultaneously inhibiting the growth of other microorganisms. Balance amongst sensitivity and selectivity is the reason for different methods for sample processing (drinking water *vs* highly contaminated waters) [28] and selection of different approaches for quantification using traditional culture based methods. The three most widely used culture based methods for quantification of FIB in any water type are Defined Substrate Technology (DST, e.g. IDEXX Colilert-18®), multiple tube fermentation (MTF), and membrane filtration (MF). The first edition of "Standard Methods for the Examination of Water and Wastewater" was released in 1905 and *E. coli* was selected as the most suitable indicator organism for raw drinking water [29]. Therefore multiple detection methods for *E. coli* were available early in the century. As scientific knowledge evolved a broader group of microorganisms (still including total coliforms, fecal coliforms, and *E. coli*) have been selected as surrogate measures in a wide range of water sources and treated water [30]. Two key factors have led to the trend of using *E. coli* as the 'preferred' indicator for the detection of fecal contamination, not only in drinking water but in other matrices as well: first, the finding that some 'fecal coliforms' were non-fecal in origin, and second, the development of improved testing methods for *E. coli* [1-19]. Membrane filtration is commonly selected for FIB quantification along the drinking water treatment process since it is flexible to the amount of sample filtered and since the method permits specific quantification. That is, both MTF and DST-based methods are either reported in a presence/absence format or in a "most probable number" format. Colilert® and Colilert- 18®, however, tend to be used at the "end of pipe" and are user friendly methods, such that their popularity has increased in the past decade.

An example of current regulations for protection of public health in Europe, USA and other countries is summarized in Table 1 [20]. Note that many currently used methods both total coliform and *E. coli*enumeration together, as IDEXX DST kits such as Colilert-18® permit the simultaneous quantification of both groups. Furthermore, while guidelines have been issued in different countries, as highlighted in Table 1, the interpretation and implementation of monitoring typically falls in the hands of member states and provinces; this creates a wide range of monitoring approaches that are currently employed for management of drinking, recreational, and shellfish harvesting water quality. It is for this reason that only a table for drinking water is presented, as drinking water regulations are the most stringent and consistent across nations.

Table 1. International drinking water standards and guidelines from the World Health Organization (WHO) Modified from [31, 32 and 33]

Parameter	Canada	United States	United Kingdom	EU directive	WHO*	Australia
Total Coliforms	0/100mL in 90%	0/100mL in 95%	0/100mL	0/100mL	*	*
Thermotolerant coliforms (fecal coliforms) or *E. coli*	*	0/100mL	*	*	0/100mL	*
E. coli	0/100mL	0/100mL	0/100mL	0/100mL	0/100mL	*
Enterococci	*	*	0/100mL	0/100mL	*	*
Cryptosporidium parvum	*	99% removal or inactivation	<1oocyte/ 10L	*	*	*
Cryptosporidium perfringens (including spores)	*	*	0/100mL	*	*	*
Pseudomonas aeruginosa	*	*	*	0/250mL	*	*
Colony count 22°C	*	*	No abnormal change	No abnormal change	*	*
Colony count 37°C	<500CFU/mL b	<500CFU/mL b	No abnormal change	20/mL c	*	*
Microcystins _LR	1,5µg/L	*	*		1,0 µg/L (provisional)	Lifetime exposure 1,3µg/L

* spaces left in blank indicate parameters not specified; a Necessary only if the water originates from or is influenced by surface water;

b HPC(35ºC for 48h) or <200 background coliforms on a total coliform membrane filter; c Necessary only in the case of water offered for sale in bottles or container

The World Health Organization (WHO) Guidelines for Drinking-water Quality [32] represent an overall international scientific consensus, based on a wide range of participants, of the health risks presented by microbes and chemicals in drinking water. The existence of international guidelines for cyanobacteria microcystins (1 µg/L microcystins –LR) in drinking water has been established by WHO (1998). Most countries rely upon this value and developed specific systems adapted for their own reality. For instance, in Brazil federal legislation requires more intensive monitoring programs including toxin analyses or toxicity testing if cyanobacteria exceed 10,000 cells/ml or 1 mm^3 biovolume in a given water sample. This includes a mandatory standard of 1 µg/L applied for microcystins (variants not specified), and recommendations limiting to saxitoxin 3µg/L.

Evaluation of water quality for drinking water production requires sampling collection at specific temporal intervals to ensure safety for the public. Usually it is recommended to sample across the process of drinking water treatment, permitting an analysis of bacteriological water quality throughout processing and distribution [6,7]. This permits rapid identification of a process or component in the distribution system that is failing, and rather than troubleshooting across the entire process, subsequent attention can be focused on specific components. This approach simultaneously prevents local contamination from developing undetected at cross-connections or breaks in the distribution lines or due to a drop in positive pressure [1, 34, 35].

The definition of a recreational water body represents a wide range of environments that include the ocean, hot springs, lakes, reservoirs, streams and rivers [34]. Similarly, the diversity of potential human pathogenic microorganisms includes all viruses, protozoans, and bacteria that could potentially be present in natural fresh and marine recreational waters particularly those contaminated by wastewater [9, 36]. In recreational water quality monitoring, monitoring approaches are often guided by use. Beaches that are used intensely by the public (such as those in Santa Monica Bay, California, USA) are monitored daily throughout the summer bathing season. In some states or areas with lower beach usage, during specific "off seasons", or areas that are used only for secondary contact recreation such as kayaking and sailing, monitoring for FIB might be conducted only once per month. Although a broad discussion is beyond the scope of this document, the use of FIB for monitoring shellfish harvesting waters also relies heavily on the use of traditional culture-based methods such as MF and MTF, but management differs by country. In some EU states, management of shellfish is by quantities of *E. coli*in shellfish meats (e.g. oysters), whereas, quantification in the United States is based upon fecal coliforms in the water column surrounding the shellfish beds.

Advantages connected to the application of culture-based methods include low equipment requirements, user-friendliness, and low cost, which makes it the most used approach for quantifying FIB in any water type. On the other hand, the enrichment stage takes time (typically from 18-24 hours) causing the notification of the public to be delayed during a contamination event [37]. This delay also causes vital revenue to be lost because contaminated waters are often still restricted or "closed" for use (in the case of beaches) after the contamination event has already passed. Incubation methods also suffer from the fact that they can promote the growth of false positive organisms [38]. Rapid molecular methods have recently been developed for application to protection of public health in recreational waters [39]. These methods require 3 hours from sample processing to results, and they are qPCR based. They target specific organisms (*Enterococcus* sp. or *E. coli*). The use of such methods permits rapid public notification (e.g. [40]) resulting in improved protection of public health. One other advantage of such methods is the high specificity [28X]. Other rapid methods using qPCR have already been developed for quantification of microbial markers that are specific to sources of fecal contamination [44, 45].

3. NEWLY DEVELOPED RAPID MOLECULAR METHODS FOR RECREATIONAL WATER QUALITY MONITORING

Millions of dollars are expended each year to measure FIB to and assess whether swimming at recreational beaches posed a risk to public health. The monitoring programs are typically conducted using culture-based methods that require an 18 to 96 hour incubation period. The lengthy incubation step results in situations where beaches are managed inaccurately, by keeping beaches open that are contaminated and closing beaches that are potentially clean again [40]. Newer, more rapid methods, such as qPCR have been developed by EPA and others for use in recreational water quality monitoring [39, 41]. Several of these new rapid PCR-based methods for quantification of *Enterococcus* sp. have been found to be significantly related to human health outcomes in epidemiology studies [42, 43]. In some cases, the association between rapid methods and adverse human health outcomes has been stronger than that observed between traditional culture-based methods and adverse human health outcomes [43].

Quantitative PCR is a novel primer-based molecular technique that combines the specificity of traditional PCR with the quantitative measurement of fluorescence for quantification of presence of specific types of nucleic acid in environmental samples. Many of the recently developed methods are actually quite similar to one another; differing only in the type of qPCR fluorophore "chemistry" used [39,41]. Even though epidemiological studies to assess relationships between rapid FIB quantification methods and human health outcomes have been successful, the implementation or widespread use of the rapid qPCR-based methods has not been without hurdles. Noble et al., 2010 [41] successfully demonstrated the use of rapid molecular methods by water quality personnel that had no previous experience with molecular techniques. Griffith and Weisberg, 2011 [40] built upon this success, by training three water quality laboratories to conduct the rapid methods, and implementing the use of the rapid methods for active beach management decisions (posting and closing of beaches). They also demonstrated the use of real time notification technology at the beaches to rapidly convey real-time water quality results through the assistance of a non-profit organization called "MiOcean". Griffith and Weisberg (2011) reported excellent management agreement (>96%, i.e. the agreement in whether a beach would be posted or remain open) among the rapid QPCR based enumeration of *Enterococcus* sp. [41] and culture based methods such as EPA Method 1600 (membrane filtration based quantification for *Enterococcus* sp.). Lavander and co-workers [46] have shown similar successes with quantification of *E. coli* at freshwater Great Lakes beaches, reporting management agreement of 98% between the rapid qPCR based quantification of *E. coli* and traditional culture based methods such as Colilert-18.

Even though success has been demonstrated in specific environments, it is clear that rapid qPCR-based methods for quantification of FIB are not appropriate for all places at all times. Three major issues to be resolved are ways to standardize assays, the approach used for quantification of the target organism, and inhibition of the qPCR.

Currently there are two modes of quantification being employed in the literature, both with advantages and disadvantages. One relies on quantifying the target cell, e.g. *Enterococcus* sp., where results are reported as cell equivalents/100 ml, where inhibition is assessed using a specimen processing control, but the final reported target concentrations are not modified quantitatively using inhibition data (see [40] for an example). The second approach reports target cell concentrations as calibrated cell equivalents/100 ml and relies on the use of cycle threshold ratios for both the target cell and the specimen processing control for final calculated quantities of the target cell concentration. The cell equivalent approach can, in certain circumstances, provide data that is more related to historical data reported for culture based methods [40,41, 46], but cannot be used at beaches were partial inhibition of the qPCR is common. On the other hand, the calibrated cell equivalent quantification approach can be more widely applied at a range of beaches, because inhibition is accounted for. However, the approach is imperfect because the specimen processing controls that are currently used are DNA-based, and they do not adequately predict inhibition of the amplification of FIB cells in real world samples [46].Dilution is another approach that can be used to effectively reduce inhibition, but this increases the limit of detection.

Figure 1 is a snapshot of results from a preliminary assessment of the performance of rapid qPCR-based methods as compared to membrane filtration (culture-based) methods for the quantification of Enterococcus sp. in water samples collected from an array of beach locations in Santa Monica Bay, California, USA. The rapid methods used here provide results within a few hours of sample collection. Quantification was conducted using the cell equivalent approach.

Inhibition of the qPCR has been discussed at length in other publications but is relevant to the implementation of rapid qPCR based methods for water quality assessment because of its unpredictable nature. Inhibition of qPCR can occur when high molecular weight compounds in the surface water (e.g., humic acids and other complex carbohydrates) combine with metal ions to sequester nucleic acids from polymerases and prevent amplification [47, 48, 49, 50]. Inhibition can sometimes be alleviated with the use of a commercial DNA extraction kit for sample purification. This requires extra time that is added to the total analysis time, and increases variability in the final results due to analyst error and variable DNA binding characteristics. Others have tried to deal with inhibition in analysis of surface and other water samples with the addition of adjuvants (e.g. bovine serum albumin, [49]. In recreational waters, many use salmon testes DNA to assess inhibition (e.g. 39). Cao et al. 2012 [10] assessed the use of dilution, salmon testes DNA, and internal controls for the assessment of inhibition in a range of water sample types, and no one approach clearly emerged as superior over the others. In fact, the internal controls and salmon testes DNA did not agree in their assessment of inhibition of the target qPCR [52]. It may be that implementation of the rapid methods, at least at first, follows the approach developed in southern California by [40].

They utilized the cell equivalent quantification approach, and for beaches that demonstrated inhibition of the QPCR, traditional culture based methods were utilized to quantify FIB concentrations for beach management purposes.

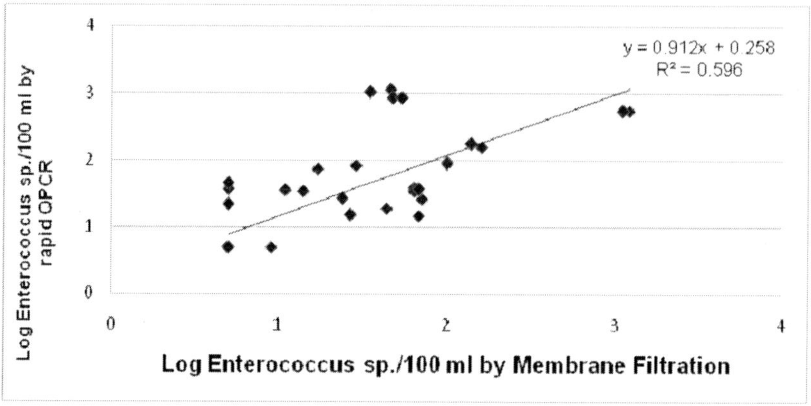

Figure 1. Example of rapid qPCR-based methods compared to membrane filtration (culture based) methods for the quantification of Enterococcus sp. in water samples collected from an array of beach locations in Santa Monica Bay, California, USA.

In the past years another new, rapid and reliable methodological approach has been tested to monitor microbial contamination, particularly in recreational marine and freshwater sources [53, 54]: the immunomagnetic separation / adenosine triphosphate (IMS / ATP) method. IMS-ATP was first developed by [55] and consists of the use of immune magnetic beads that selectively capture targeted bacteria using cell specific antibodies. A magnet is then used to separate the target bacteria from the rest of the sample with concentration determined by quantification of the ATP concentration in the sample using a bioluminescent assay [53]. While IMS/ATP has not been the subject of potential recommendation by the USEPA in 2012, there are advantages to its use. These advantages include the detection of only viable bacteria, therefore making it more directly related to quantification based upon other currently used culture based methods. Rapid quantification using IMS/ATP might be beneficial in areas receiving treated, disinfected wastewater for example, where the concentration of FIB-specific DNA is likely high, but the concentration of viable FIB cells would be lower. The capital costs of IMS/ATP are lower than those of the molecular methods. Another positive aspect of this technique is that it requires less than 4 hours. The major criticism of IMS/ATP concerns the availability of appropriate antibodies for quantification, particularly for the *Enterococcus* genus, which cannot be adequately captured using a single set of antibodies. However, improvements are being made in the development of cocktails of antibodies combined with the development of new bead based capture systems [e.g. 54].

In the coming years, there is the potential to dramatically improve the sensitivity, accuracy, precision, and speed of the QPCR assays used for water quality applications, while dramatically reducing the cost associated. One key step forward will be in the area of standardization, as standardized kits and cell standards and controls become available in ready-made kits, the analyst error and variability associated with the assays will decrease dramatically. The idea is to develop an elegant and powerful, yet simple, tool that can be applied to a wide variety of water sample types, to simultaneously determine the FIB concentration and enumerate bacterial tracers or pathogen types, necessary for preliminary source identification and/or protection of public health. Many key groups around the globe are working collaboratively to ensure this success.

4. HARMFUL ALGAE BLOOMS (HABS) IN DRINKING WATER TREATMENT PLANTS AND RECREATIONAL WATERS

In general, harmful algal blooms (HABs) are a cause of great concerns amongst stakeholders involved in water quality regulation and management. Since this chapter is specifically devoted to the problems caused by microbial contamination, we are specifically focusing our treatment of the subject on the closely related cyanobacteria, and their ability to form toxic blooms (referred to herein as CYANOHABs). There is an array of reports on the increased incidence of toxic blooms caused by Cyanobacteria. Their cyanotoxins are a serious concern for raw water sources that are exploited for drinking water production and recreational purposes [33]. Most countries around the world rely upon the exploration of surface freshwater sources for drinking water production. Efficient surface water treatment systems usually combine a series of treatment stages. Commonly employed water supply treatment includes multiple steps, generally including in order: pre-chlorination, flocculation and coagulation (addition of salts of iron or aluminium), sedimentation and post-chlorination [R]. One specific problem associated with toxic blooms of cyanobacteria is that specific stages of drinking water treatment are needed to remove the toxins. Not all of the drinking water treatment plants have these stages available or possess a flexible system [56]. The goal is to guarantee the production of drinking water according to national water quality guidelines, including microbial and chemical parameters that could potentially promote human disease and/or fatalities. In Table 2 some of the international guidelines for the presence of cyanobacteria toxins is presented. Cyanotoxins can be classified from to a toxicological perspective: hepatotoxins, cytotoxins, neurotoxins, irritant compounds as well as dermatotoxins [33]. Common methods of detecting and quantifying cyanotoxins include Enzyme-linked immunosorbent assay (ELISA), and chromatographic analysis [33, 57].

The main concern for public health associated with cyanobacterial blooms is exposure to cyanotoxins which can cause central nervous system or liver damage. There are multiple modes of exposure to cyanotoxins but most common are: ingestion of contaminated water, accidental ingestion by swimming on contaminated waters, and consumption of vegetables irrigated by contaminated waters or even inhalation (Table 3) [33,57]. A CYNOHAB's cause adverse effects to drinking water production plants. For instance there are physical impacts and the release of hazard by-products from the treatment stages such as chemical treatment. Other hazards come from oxygen depletion or pipe corrosion due to microbial activity within the pipes coming from bloom collapse. It is fundamental to have knowledge of source waters such as lakes that are prone to the development of toxic and non-toxic cyanobacterial blooms so that the overall population can be informed about public health risks. Several statistical approaches have been used to create models that could explain phytoplankton feed-back to nutrient enrichment and oxygen reduction, and these models may be adapted to other locations [56,58].

Whilst WHO's guidelines are based on total of enumerated cells (using microscopy techniques) and/or determination of chla (chlorophyll a) concentrations related to the amount of cells, there is a clear desire to develop faster quantification of cyanotoxins in environmental samples, especially large population areas. It is important to have rapid, reliable and accurate analysis i.e. high throughput molecular detection methods. In order to develop molecular assays, pure cultures from toxin producers need to be utilized to focus on genes specifically involved in the synthesis of a given toxin [AS].InTable 2, the guidelines from WHO for cyanobacterial levels and expected health effects due to exposure of contaminated recreational waters are shown.

Recently published molecular approaches for detection and quantification of toxigenic and non-toxigenic cyanobacteria are based on the application of multiplex approaches for targeting the genus (16S rDNA), and subsequently using other nested assays for sequences found in the non-ribosomal polypeptide synthetize complex (55KB in size) responsible for toxin production [56,57, 58]. Al-Tabrineh et al. (2012) recently developed and validated a quadruple target qPCR assay specifically designed for the detection and quantification of genes involved in the production of microcystins, nodularins, cylindrospermopsins and saxitoxins. There is a multiplicity of qPCR protocols published recently; however they typically focus on the quantification and detection of *Microcystis* spp. as a target due to the fact that it represents the most recurrent blooming cyanobacterial group in multiple water sources around the globe [59].

CYANOHAB related problems begin at the raw water source. Therefore, international guidelines such as the Water Frame Directive (Water Frame Directive, WFD; Directive 2000/60/CE of 23 October 2000) require a good ecological state in freshwater or highly modified freshwater sources (example water dams). Therefore, if a water body doesn't comply with these requirements it should be considered contaminated treated accordingly adding necessary steps to guarantee good finished water quality [60]. Thus, regular monitoring of freshwater sources for both recreational and drinking water purposes are necessary. Current modes of quantification starts with chla, but can extend to molecular quantification of specific phytoplankton groups, quantification of toxins (typically ELISA based), or pigment based quantification of phytoplankton groups [61].

5. THE APPLICATION OF MICROBIAL SOURCE TRACKING IN WATER

5.1. Microbial Source Tracking In Water

Microbial source tracking (MST) is the use of microbial markers (including bacteriophage, bacteria, viruses, protozoans, etc.) to determine the source of fecal pollution present in an aquatic system (i.e., human, pet, livestock). Some refer to the scientific discipline as bacterial source tracking (BST), but this term is older and has given way to MST, acknowledging the wider array of information that can be gleaned from targets other than bacteria. The MST field continues to advance rapidly and much that has been published ten years ago is now obsolete and so it is generally better to refer to studies conducted since 2005 for the most updated information. For preliminary information on the evolution of the MST field, the reader should refer to the US Environmental Protection Agency's MST Guide (EPA 2005), article by [62]. MST methods were once categorized into two types — library dependent approaches and library independent approaches. Library dependent methods relied on matching the known source "fingerprints" or isolate patterns generated by molecular or phenotypic methods to unknowns collected from the environment. The most commonly used library dependent methods include, antibiotic resistant analysis (ARA), carbon utilization profiles (CUP), pulse field gel electrophoresis (PFGE), restriction- or amplified- fragment length polymorphism (RFLP or AFLP), random amplified polymorphic DNA (RAPD), repetitive extragenic palindromic PCR (rep-PCR) and ribotyping.

Library independent approaches are based upon detection (often presence/absence) or quantification of specific markers of fecal contamination, and are not dependent upon any sort of classification into categories. Library independent methods include F-specific and somatic coli phage serotyping and genotyping, and host-specific and marker-specific PCR and qPCR for a wide range of targets (see Table 3). Many older approaches utilized conventional PCR, but those have generally given way to the increased use of qPCR. Conventional PCR results are typically reported as presence/absence, or at best by using a serial dilution technique, can produce semi-quantitative information [63]. As opposed to conventional PCR, which is limited to a presence/absence result, qPCR provides for quantification over a wide dynamic range. The platforms currently being used for qPCR include hardware with proven rapid cycling, sensitive optics, and multiplex capabilities.

Table 2. WHO guidance values for recreational waters with respect to cyanobacteria contaminates waters.(modified from [33]).

Guidance level or situation	How guidance level derived	Health risks
20000 cyanobacterial cells/mL 10µg chlorophyll-a/L Low impacts in human health	From human bathing epidemiological study	Skin irritations and gastrointestinal illness
100000 cyanobacterial cells/mL or 50µg chlorophyll-a/L moderate chances of affecting human health	From provisional drinking water guideline value and data concerning other cyanotoxins	Long-term illness and/or Short-term problems such as skin irritations and gastrointestinal illness
High probability of adverse effects Scum formation in areas where whole-body contact and/or risk of ingestion/aspiration occur	Inference from oral animal lethal poisonings Actual human illness case histories	Potential for acute poisoning Potential for long-term illness with Adverse health problems such as skin irritations and gastrointestinal illness

Table 3. Examples of cyanobacterial toxins and their acute toxicity modified from [33]

Cyanotoxins	Mechanism of toxicity	Taxa toxin producer	LD_{50}(mouse) pure toxin
Protein-phosphatase blockers			
General microcystins	Blocks protein phosphatases by covalent binding promoting liver haemorrhage damage could occur	Microcystis, Planktothrix, Anabaena, Oscillatoria, Nostoc, Anabaenopsis, Hapalosiphon, Nodularin	45->1000µg/Kg
Microcystin-LR			60(25-125)µg/Kg
Microcystin-YR			70 µg/Kg
Microcystin-RR			300-600 µg/Kg
Nodularin			30-50 µg/Kg
Neurotoxins			
Anatoxin-a	Blocks postsynaptic depolarization	*Aphanizomenon, Oscillatoria, Anabaena, Cylindrospermum*	250 µg/Kg
Saxitoxins	Blocks sodium channels	*Aphanizomenon, Anabaena, Lyngbya, Cylindrospermopsis raciborskii*	10-30 µg/Kg
Cytotoxin			
Cylindrospermopsin	Blocks protein synthesis; substantial cumulative toxicity	*Cylindrospermopsis raciborskii*	200 µg/Kg/d

5.2. Bacteroides Based Microbial Source Tracking

One of the most fruitful organisms of focus in MST has been the Bacteroidales family. The Bacteroidales family has yielded a range of quantitative markers for human, dog, cow, horse, and other types of fecal contamination (e.g. 23, 71) The Bacteroides genus was suggested as early as 1967 as an alternative indicator due to the fact that *Bacteroides* spp. levels present in human and animal intestinal tracts are about 10 to 100 times above the other FIB [14, 16]. The high degree of difficulty for cultivation of *Bacteroides* spp. reduced its use in the late decades of the 20th century, but the development of molecular tools allowed the detection and quantification of members of the family Bacteroidales, without cultivation [64]. Consequently, there has been a dramatic increase in publications detailing development of qPCR assays for the quantification of Bacteroidales in recreational waters as indicators of specific types of fecal contamination [14,65]. While several studies have demonstrated the value of the family Bacteroidales, some have also highlighted problems associated with cross-reactivity for certain markers, making testing and validation of assays particularly important [64, 65]. One of the most important advancements has been the development of the HF183 human specific markers that has been shown in a range of studies to be specific and relatively sensitive [66,67,68]. Acknowledging that no one marker can be used everywhere and in all circumstances, the HF183 marker has been atypical is that it has been successfully utilized more than any other single marker in application for detection and quantification of human fecal contamination in recreational waters. Recent work [69] using the HF183 marker has shown that data generated can be suitable for providing stakeholders and decision markers with valuable information for remediation projects [31].

Recently, another qPCR-based method has been developed by [70] to quantify fecal *Bacteroides* spp. The assay was specifically designed to select for the *Bacteroides* species that are most human-associated, but it also amplifies Bacteroides spp. from animals. The focus of the assay is on a cohort of*Bacteroides* species that are most prominent in the human gut. This method has been successfully used as a screening tool, with more specific methods used to quantify specific types of fecal contamination [71]. Another successful approach is reported by [71], where the authors developed an array of fecal*Bacteroides* based markers for a range of types of fecal contamination. They reported a quantitative universal *Bacteroidales* assay called BacUni-UCD that detected universal fecal *Bacteroidales*in all of the test samples (n=73) including human, cat, seagull, horse, dog and cow feces. The human assay BacHum-UCD presented by [68], successfully discriminated fully between human and cow stool samples and had slight cross-reactivity with dog stool, however, the method is highly quantitative when used in combination with human specific approaches outlined by [68]. All of the wastewater samples tested during the [68], study were positive for the BacHum-UCD marker, showing it to be 100 percent sensitive for human source identification. In 2009, Ahmed and co-workers [73] published a test of the host specificity of the five published sewage-associated *Bacteroides* markers. They tested 186 fecal samples from 11 animal species including humans. All of the human fecal samples were positive for all five markers indicating 100 percent sensitivity of the above referenced markers of [71], [68] and [72]. The HF183 marker has been found to have the capability to differentiate between human and animal feces with 99 percent accuracy [73]. It has been used in several mid-Atlantic stormwater studies to indicate the presence of human fecal contamination [74].

5.3. Virus Based Methods

Pathogen-based research for MST has focused almost solely upon pathogens that are carried via the fecal-oral route. The work that has been accomplished regarding enteric viruses has relied on the fact that high concentrations of these viral pathogens can be found in human sewage. High concentrations are a vital characteristic, given the limitations with concentration and isolation of pathogens in wastewater and complex aquatic matrices. In 2005, in a review written by Fong and Lipp [75], the authors recount the major developments in the past decades, including applications of human and animal viruses to source tracking. The authors detail the disadvantages and advantages of virus-based detection and quantification methods. In particular, they present a summary of information of the myriad of animal viruses applied to source tracking studies for human fecal contamination, namely human enteroviruses (echovirus and Coxsackie virus for example) and adenoviruses (Ad5, 7, 40, and 41). The review is completed by iterating the need for strong study designs with the use of virus detection and quantification methods for microbial source tracking applications, due to the limitations with the ability to detect viral targets, which can often occur in aquatic systems at low concentrations and high patchiness.

Rajal and co-workers [76] presented work on the quantification of human and animal pathogenic viruses in real world water samples. They used a known sample volume and tracer addition of surrogate viruses to calculate the limit of detection. This work was one of the first to detail carefully the quantification and recovery in a tightly designed study. For any type of human pathogenic virus quantification, recovery estimations are vital. This can typically be done through the use of spiked samples or by adding surrogate DNA or RNA to the sample, or with the use of a competitive internal positive control (CIPC). In 2006, Gregory et al.[78] reported the first use of a CIPC for full quantification of RNA viruses for water quality applications. The point of this internal control was to permit assessment of the efficiency of the combined reverse-transcriptase and PCR steps to enterovirus quantification and to provide a scenario for CIPC design approaches for other virus types. Similar approaches were developed by [79] and have resulted in successful quantification of important human pathogenic enteroviruses in recent years.

While not all viruses will be infectious in a water sample, accurate quantification of viruses in a water sample along with flow measurements can permit loading estimates of potentially pathogenic material, specific to source, thereby permitting partitioning specific sources of fecal contamination in complex environments. Also, the accurate quantification of the human pathogenic viruses is an important parameter in the successful determination of the potential public health risk and can be a key component to Quantitative Microbial Risk Assessment (QMRA).

Recently, human polyomaviruses (HPyV) have emerged as a useful human-specific viral marker. There has been a series of studies conducted that report the presence and quantification of these viruses in raw sewage samples and/or environmental waters in sites ranging from Florida [82], California [82,83], Australia [66], Spain [84], Germany [85] and Japan [86]. These viruses are double-stranded DNA viruses frequently isolated from urine, and in some cases feces, of both healthy and immunocompromised individuals. In a recent study conducted by [87], both nested PCR and qPCR were used to assess the presence of HPyV in raw and treated sewage in Rio de Janeiro, Brazil. The study interestingly demonstrated the clustering of different types of the HPyV coming from distinct African and European lineages. While this study was focused on samples from a sewage treatment plant, it demonstrates the specificity of information that can be gleaned from viral pathogen-based investigations.

A very recent publication by [87] has demonstrated the use of a range of MST approaches at a beach in southern California that has been the site of a range of water quality investigations (e.g. [43, 88]). Doheny State Beach, in Orange County, California, USA is a high priority, intensely utilized beach location that has suffered from a range of beach water quality problems, In the past decade, for example, the beach has been given a poor grade several times on the "Heal the Bay Beach Report Card" (www.healthebay.org). This study demonstrated the utility of the HF183 human specific marker, adenovirus, HPyV, and *Methanobrevibacter smithii* as markers of fecal contamination. They found distinct correlations among the highly human specific molecular markers, i.e. adenovirus at Doheny State Beach was correlated to both HPyVs and the HF183 molecular marker. This study dramatically presents the benefits of application of a range of microbial source tracking approaches (i.e. the use of the toolbox approach), as opposed to a single marker. It also demonstrates the complexity of real-world water quality investigations and the importance of conducting studies over appropriate time and sampling scales.

6. CONCLUSIONS

Over the past decades, the use of culture-based FIB based quantification approaches has dominated the management of microbial contaminants for drinking water, recreational water, and shellfish harvesting water quality. While these approaches are useful and have resulted in a dramatic improvement in water quality management over that observed prior to the 1950's, there is an array of new molecular approaches that can be used to bolster water quality management and the process of public notification (i.e. water advisories, beach posting and closings, and shellfish harvesting water closures).

Here, highlights have been presented of three areas for immediate advancement of the science. Rapid qPCR based methods are poised for future use to improve the accuracy of beach water quality postings and beach closures. Molecular assays specific for toxin producing strains of cyanobacteria, and other HAB species can be utilized to improve the management of our vital raw surface waters. Microbial source tracking techniques are poised to improve all facets of water quality management; from site specific investigations of the predominant fecal sources to assessment of the efficiency of wastewater treatment systems.

This is an exciting time for researchers given the fast evolution of molecular approaches. The design of appropriate methods, that are versatile, fast and reliable, can permit information to be transmitted to managers in shorter periods of time. The molecular methods will only improve in their cost effectiveness and user-friendliness in the next few years. Finally, the successful application of predictive multivariate models and formalization of QMRA approaches will improve dramatically with the infusion of the scale appropriate molecular information on microbial contaminants, sources of contamination and rates of discharge and loading.

REFERENCES

1. Tallon P, Magajna B, Lofranco C and Leung KT. Microbial indicators of fecal contamination in water: a current perspective. Water, Air, and Soil Pollution, 2005; 166:139–166

2. Bláha L, Babica P, Blahoslav. Toxins produced in cyanobacterial water blooms – toxicity and risks Interdisc Toxicol, 2009; 2(2): 36–41

3. Standard methods for the examination of water and wastewater. Greenberg, A.E, Clesceri, L.S. and Eaton, A.D. (Eds). 18th edition; 1992;

4. Balarajan R, Raleigh VS, Yuen P, Wheeler D, Machin D, Cartwright R.(1991) Health risks associated with bathing in seawater. British Medical Journal, 1991; 303: 1444

5. Haile RW, Witte JS, Gold M, Cressey R, McGee C, Millikan RC, Glasser A, Harawa N, Ervin C, Harmon P, Harper J, Dermand J, Alamillo J, Barrett K, Nides M, Wang GY. The health effects of swimming in ocean water contaminated by storm drain runoff. Epidemiology, 1999; 10: 355-363

6. Dufour AP, Ballentine P. Ambient water quality criteria for bacteria. Technical Report EPAA440/5-84-002. U.S. Environmental Protection Agency, Washington DC, 18, 1986

7. Izbicki JA, Swarzenski PW, Reich CD, Carole Rollins C and Holden, PA. Sources of fecal indicator bacteria in urban streams and ocean beaches, Santa Barbara, California. Annals of Environmental Science, 2009; 3:139-178

8. World Health Organization. Report of WHO Scientific Group of human viruses in water, wastewater and soil. Technical Report Series, no. 639. World Health Organization, Geneva, Switzerland, 1979

9. Ikner LA, Gerba CP and Bright KR,. Concentration and Recovery of Viruses from Water: A Comprehensive Review on Food Environmental Virology, 2012;4:41–67

10. Cao Y, John F Griffith JF, Samuel Dorevitch S and Stephen B Weisberg SB. Effective- ness of qPCR permutations, internal controls and dilution as means for minimizing the impact of inhibition while measuring Enterococcus in environmental waters. Journal of Applied Microbiology 2012; 113(1):66-75

11. World Health Organization. Report of WHO Guidelines for drinking water quality, I-recommendations. World Health Organization, Geneva, Switzerland, 2004;

12. Figueras MJ and Borrego JJ.. New Perspectives in Monitoring Drinking Water Microbial Quality. Int. J. Environ. Res. Public Health, 2010; 7:4179-4202

13. U.S. ENVIRONMENTAL PROTECTION AGENCY. Ambient Water Quality Criteria for Bacteria. EPA-440/5-84-002, U.S. Environmental Protection Agency, Washington, D.C., 1986

14. Griffith JF, Cao Y, McGee CD, Stephen B. and Weisberg B. Evaluation of rapid methods and novel indicators for assessing microbiological beach water quality. Water Research, 2009; 43:4900 – 4907

15. Harmsen, H J. M, Raangs GC, He T, Degener JE, and Welling GW. Extensive set of 16S rRNA-based probes for detection of bacteria in human feces. Applied and Environmental Microbiology, 2002; 68:2982–2990

16. Leila Kahlisch I, Henne K, and Gröbe, L.. Assessing the Viability of Bacterial Species in Drinking Water by Combined Cellular and Molecular Analyses. Microb Ecol 2012; 63:383–397

17. Eichler S, Christen R, Höltje C, Westphal P, Bötel J, Brettarl,Mehling A. and Höfle MG. Composition and dynamics of bacterial communities of a drinking water supply system as assessed by RNA- and DNA-based 16S rRNA gene fingerprinting. Applied and Environmental Microbiology, 2006; 72:1858–1872

18. Paavola, J., Institutions and environmental governance: a reconceptualization. Ecological Economics 2007; 6, 1:93–103

19. Orin C. Shanks, O C, Kelty,CA, Sivaganesan, M, Varma M and Richard A. Haugland, RA Quantitative PCR for Genetic Markers of Human Fecal Pollution, Applied and Environmental Microbiology, 2009;.75, 17:5507–5513

20. U.S. ENVIRONMENTAL PROTECTION AGENCY, Ambient Water Quality Criteria for Bacteria. EPA-440/5-84-002, U.S. Environmental Protection Agency, Washington, D.C., 1986

21. Dafour, AP. Health Effects Criteria for Fresh Recreational Waters. EPA-600/1-84-004, U.S. Environmental Protection Agency, Research Triangle Park, N.C., 1984

22. Davies, CM, Long, JA, Donald, M and N. J. Ashbolt, NJ, Survival of fecal microorganisms in marine and freshwater sediments. Applied and Environmental Microbiology 1995; 61:1888–1896

23. Layton, A., McKay L,Williams D, Garrett V, Gentry, R and Sayler. G. Development of Bacteroides 16S rRNA gene TaqMan-based real-time PCR assays for estimation of to- tal, human, and bovine fecal pollution in water. Applied Environ Microbiology 2006; 72:4214–4224

24. Matsuki T, Watanabe K, Fujimoto J, Takada T, and Tanaka, R. Use of 16S rRNA gene- targeted group-specific primers for real-time PCR analysis of predominant bacteria in human feces. Applied and Environmental Microbiology, 2004; 70:7220–7228

25. Boehm, AB, Ashbolt, NJ. A sea change ahead for recreational water quality criteria. Journal of Water Health 2009; 7, 1: 9-20

26. Gentry, R W, Layton AC, McKay LD, McCarthy JF, Williams DE, Koirala SR, and Sayler GS. Efficacy of Bacteroides measurements for reducing the statistical uncertainty associated with hydrologic flow and fecal loads in a mixed use watershed. Journal of Environmental Quality, 2007; 36:1324–1330

27. Figueras MJ, Robertson W, Pike E and Borrego JJ. Sanitary inspection and microbiological water quality. In: Monitoring Bathing Waters. A Practical Guide to the Design and Implementation of Assessments and Monitoring Programs; Bartram J. and Rees G (Eds), E & FN Spon: London, UK, 2000; pp. 113-126

28. Fielda KG and Samadpourb M. Fecal source tracking, the indicator paradigm, and managing water quality. Water Research, 2007; 41: 3517-3538

29. Environmental Health Directorate – Health Protection Branch. Recreational Water Quality-Published by Authority of the Minister of National Health and Welfare, 1978;

30. Kornacki JL and Johnson JL. Enterobacteriaceae, Coliforms and *Escherichia coli* as quality and safety indicators'. In: F. Downs (Eds), Compendium of Methods for the Microbiological Examination of Foods, APHA. Washington DC, 2001;

31. World Health Organization. Guidelines for drinking-water quality, 3rd ed., vol. 1. World Health Organization, Geneva, Switzerland; 2008

32. WHO, Guidelines for Drinking-Water Quality, 2nd edition Geneva, 1993; ISBN 92 4 154460(Albinana-Gimenez et al. 2009), Germany (Hamza et al. 2009) and Japan (Haramoto et al. 2010)

33. Chorus I and Bartram J (Eds.). Toxic cyanobacteria in water: a guide to their public health consequences, monitoring and management. E&FN Spon, ISBN 0-419-23930-8, London, 1999;

34. Izbicki JA, Swarzenski PW, Reich CD, Carole Rollins C and Holden, PA. Sources of fecal indicator bacteria in urban streams and ocean beaches, Santa Barbara, California. Annals of Environmental Science, 2009; 3:139-178

35. Whitman RL, Przzybyla-Kelly K,Shivley DA and Byappanhali MN. Incidence of the enterococcal surface protein (esp) gene in human and animal fecal sources. Environ SciTechnol 2007; 41(17):6090–6095

36. Dafour, AP. Health Effects Criteria for Fresh Recreational Waters. EPA-600/1-84-004, U.S. Environmental Protection Agency, Research Triangle Park, N.C., 1984

37. Leecaster MK and Weisberg SB. Effect of Sampling Frequency on Shoreline Microbiology Assessments. Marine Pollution Bulletin, 2003;42(11): 1150-1154.

38. Pisciotta J, et al. The role of neutral lipid nanospheres in *Plasmodium falciparum* crystallization. Biochem J, 2007, 402:197–204.

39. Haugland RA, Siefring SC, Wymer LJ, Brenner KP and Dufour AP. Comparison of Enterococcus density measurements by quantitative polymerase chain reaction and membrane filter culture analysis at two freshwater recreational beaches. Water Research, 2005; 39:559-568.

40. Griffith JF and Weisberg SB. Challenges in Implementing New Technology for Beach Water Quality Monitoring: Lessons from a California Demonstration Project. Marine Technology Society Journal 2011; 45:65–73

41. Noble RT, Blackwood AD, Griffith JF, McGee CD and Weisberg SB. Comparison of Rapid Quantitative PCR-Based and Conventional Culture-Based Methods for enumeration of *Enterococcus* spp. and *Escherichia coli* in Recreational Waters. Applied and Environmental Microbiology, 2010; 76(22):7437-7443

42. Wade TJ, Calderon RL, Brenner KP, Sams E, Beach M, Haugland R, Wymer L and A. P. Dufour AP. High Sensitivity of Children to Swimming-Associated Gastrointestinal Illness Results Using a Rapid Assay of Recreational Water Quality. Epidemiology, 2008; 19: 375-383

43. Colford JM, Schiff K, Griffith JF, Yau V, Arnold BF, Wright C, Gruber J, Wade T, Burns S, Hayes S, McGee C, Gold M, Noble RT and Weisberg, SB. Using rapid indicators for Enterococcus to assess the risk of illness after exposure to urban runoff contaminated marine water. Water Research, 2012; 1-11.

44. Scott TM, Jenkins TM,, Lukasik J and Rose JB. Potential use of a host-associated molecular in *Enterococcus faecium* as an index of human fecal pollution. Environmental Science and Technology, 2005; 39 (1):283–287.

45. McQuaig SM, Scott TM, Harwood VJ, Farrah SR and Lukasik JO. Detection of hu- man-derived fecal pollution in environmental waters by use of a PCR-based human polyomavirus assay. Applied and Environmental Microbiology, 2006; 72(12):7567–7574

46. Lavender JS and Kinzelman JL. A cross comparison of QPCR to agar-based or de- fined substrate test methods for the determination of *Escherichia coli* and enterococci in municipal water quality monitoring programs. Water Research 2009; 43(19): 4967-4979

47. Tsai YL and. Olson BH. Rapid method for separation of bacterial DNA from humic substances in sediments for polymerase chain reaction. Applied and Environmental Microbiology, 1992; 479(58):2292-2295

48. De Boer SH, L.J. Ward LJ, X. Li X and S. Chittaranjan. S. Attenuation of PCR in- hibition in the 411 presence of plant compounds by addition of BLOTTO. Nucleic Acids Research, 1995; 23:2567-2568

49. Kreader CA. Relief of amplification inhibition in PCR with bovine serum albumin or T4gene 32 protein. Applied and Environmental Microbiology, 1996; 62:1102-1106

50. Watson RJ and B. Blackwell. Purification and characterization of a common soil component which inhibits polymerase chain reaction. Canadian Journal of Microbiology, 2000; 46:633-642

51. Thurman EM, Aiken GR, Ewald M, Fischer WR, Forstner U, Hack AH, Mantoura RFC, Parsons JW, Pocklington R, Stevenson FJ, Swift RS and SzpakowskaB.Isolation of soil and aquatic humic substances. pp. 31-43 in: F.H. Frimmel and R.F. Christman (ed.), Humic Substances and Their Role in the Environment. John Wiley and Sons, Ltd. New York, NY, 1988

52. Cao Y, Griffith JF and Weisberg SB. Evaluation of optical brightener photo decay characteristics for detection of human fecal contamination. Water Research, 2009; 49 (8):2273-227

53. Bushon RN, Brady AM, Likirdopulos CA and Cireddu JV. Rapid detection of *Escherichia coli* and enterococci in recreational water using an immunomagnetic separation/ adenosine triphosphate technique Journal of Applied Microbiology, 2009;106: 432–441

54. Lee CM, Griffith JF, Kaiser W and Jay, JA. Covalently linked immune magnetic sepa- ration/ adenosine triphosphate technique (Cov-IMS/ATP) enables rapid, in-field detection and quantification of *Escherichia coli* and Enterococcus spp. in freshwater and marine environments. Journal of Applied Microbiology, 2009; 109:324–333

55. Lee JY. and Deininger RA. Detection of *E-coli* in beach water within 1 hour using immunomagnetic separation and ATP bioluminescence. Luminescence 2004; 19:31–36

56. Wiegand C and Pflugmacher S. Ecotoxicological effects of selected cyanobacterial metabolites a short review. Toxicology and Applied Pharmacology, Vol. 203, (March 2005); pp. 201-218, ISSN 0041-008X

57. Al-Tebrineh J, Leanne A. Pearson LA, Serhat A. Yasar SA and Brett A. Neilan BA. A multiplex qPCR targeting hepato- and neurotoxigenic cyanobacteria of global significance. Harmful Algae, 2012; 15:19–25

58. Sivonen K and Jones G. Cyanobacterial toxins. In: Toxic cyanobacteria in water: a guide to public health significance, monitoring and management. Chorus I and Bar- tram (Eds).The World Health Organization, ISBN 0-419-23930-8, E and FN Spon, London, UK, 41-111pp, 1999

59. Conradie KR and Barnard S. The dynamics of toxic *Microcystis* strains and microcys- tin production in two hypertrofic South African reservoirs. Harmful Algae 2012; in press

60. Henderson, Chips M, Cornwell, Hitchins P, Holden B, Hurley S, Parsons SA, Wetherill A and Jefferson B. Experiences of algae in UK waters: a treatment perspective. Water and Environment Journal, 2008; 22:184–192

61. Pinckney JL, Richardson TL, Millie DF, Paerl HW. Application of photopigment biomarkers for quantifying microalgal community composition and in situ growth rates. Organic Geochemistry, 2001; 32: 585–595

62. Simpson JM, Santo Domingo JW and Reasoner DJ. Microbial source tracking: state of the science. Environmental Science and Technology 2002; 36:5279–5288

63. Noble RT and J. A. Fuhrman, JA. Enteroviruses detected in the coastal waters of Santa Monica Bay, CA: Low correlation to bacterial indicators. Hydrobiologia, 2001;460:175-184

64. Gentry, R W, Layton AC, McKay LD, McCarthy JF, Williams DE, Koirala SR, and Sayler GS. Efficacy of Bacteroides measurements for reducing the statistical uncertainty associated with hydrologic flow and fecal loads in a mixed use watershed. Journal of Environmental Quality, 2007; 36:1324–1330

65. Boehm, AB, Ashbolt, NJ. A sea change ahead for recreational water quality criteria. Journal of Water Health, 2009; 71: 9-20

66. Ahmed W, Stewart J, Powell D and Gardner T. Evaluation of Bacteroides markers for the detection of human faecal pollution. Letters in Applied Microbiology, 2008; 46(2):237-242

67. Ahmed W, Goonetilleke A, Powell D, Chauhan K and Gardner T. Comparison of molecular markers to detect fresh sewage in environmental waters. Water Research, 2009;43:4908-4917

68. Seurinck S, Defoirdt T, Verstraete W and Siciliano SD.Detection and quantification of the human-specific HF183 Bacteroides 16S rRNA genetic marker with real-time PCR for assessment of human faecal pollution in freshwater. Environmental Microbiology, 2005; 7(2):249-59

69. Mieszkin S, Furet JP, Corthier G, and M. Gourmelon M. Estimation of pig fecal con- tamination in a river catchment by real-time PCR using two pig-specific Bacteroi- dales 16S rRNA genetic markers. Applied and Environmental Microbiology 2009; 75:3045–3054

70. Converse RR, Blackwood AD, Kirs M, Griffith JF and Noble RT, Rapid QPCR-based assay for fecal Bacteroides spp. as a tool for assessing fecal contamination in recreational waters. Water Resources, 2009; 43:4828–4837

71. Kildare BJ, Leutenegger CM, McSwain SM, Bambic DG, Rajal VB and Wuertz S. 16S rRNA-based assays for quantitative detection of universal, human-, cow-, and dog-specific fecal Bacteroidales: A Bayesian approach. Water Research, 2007; 41:3701-3715

72. Bernhard A and Field, KG. Identification of nonpoint sources of fecal pollution in coastal waters by using host-specific 16S ribosomal DNA genetic markers from fecal anaerobes. Applied and Environmental Microbiology, 2000; 66:1587–1594

73. Ahmed W, Goonetilleke A, Powell D, Chauhan K and Gardner T. Comparison of molecular markers to detect fresh sewage in environmental waters. Water research, 2009; 43:4908-4917

74. Parker JK, McIntyre D and Noble RT. Characterizing fecal contamination in storm-water runoff in coastal North Carolina, USA. Water Resources, 2010; 44(14):4186-4194

75. Fong TT and Lipp EK. Enteric viruses of humans and animals in aquatic environments: health risks, detection, and potential water quality assessment tools. Micro Mol Biol Rev, 2005; 69(2):357-371

76. Rajal, McSwain BS, Thompson DE, Leutenegger CM, Wuertz, S. Molecular quantitative analysis of human viruses in California stormwater. Water Research, 2007; 41(19):4287-4298

77. Wong M, Kumar L, Jenkins TM, Xagoraraki I, Phanikumar MS and Rose JB, (2009) Evaluation of public health risks at recreational beaches in Lake Michigan via detection of enteric viruses and a human-specific bacteriological marker. Water re- sources, 2009; 43:1137-1149

78. Gregory JB, Litaker RW and Noble RT. Rapid One-Step Quantitative Reverse Tran- scriptase PCR Assay with competitive internal positive control for detection of En- teroviruses in environmental samples. Applied and Environmental Microbiology, 2006; 72(6):3960-3967

79. Fuhrman JA, Liang X and Noble RT. Rapid Detection of Enteroviruses in Small Vol- umes of Natural Waters by Real-Time Quantitative Reverse Transcriptase PCR Ap- plied and Environmental Microbiology, 2005; 71(8): 4523–4530

80. Baums L, Goodwin KD, Kiesling T, Wanless D, Diaz MR, Fell,]W. Luminex detection of fecal indicators in river samples, marine recreational water, and beach sand. Marine Pollution Bulletin, 2007; 54 (5):521-536

81. Carson CA, Christiansen JM, Yampara-Iquise H, Benson VW, Baffaut C, Davis JV, Broz RR, Kurtz WB, Rogers WM and Fales WH. Specificity of*Bacteriodes thetaiotao- micron* marker for human feces. Applied and Environmental Microbiology, 2005; 71(8):4945-4949

82. McQuaigSM , Scott TM , Lukasik JO , Paul JH , Harwood VJ . Quantification of human polyomaviruses JC virus and BK virus by TaqMan quantitative PCR and comparison to other water quality indicators in water and fecal samples .Applied and Environmental Microbiology 2009; 75: 3379-3388

83. Rafiquea and Jiang, CS. Genetic diversity of human polyomavirus JCPvV in Southern California wastewater. Journal of Water Health, 2009; 6:533-538

84. Albina-Gimenez N, Clemente-Casares P, Calgua B, Huguet, JM, Courtois S and Gi- rones R. Comparison of methods for concentrating human adenoviruses, polyomavi- rus JC and noroviruses in source waters and drinking water using quantitative PCR. J Virol Methods, 2009; 158: 104-109

85. Hamza IA, Jurzik L, Stang K, Uberla K and Wilhem M. Detection of human viruses of a densly-populated area in Germany using virus adsorption method optimized for PCR analyses. Water Resources, 2009; 43:2657-2668

86. Haramoto E, Kitajima M, Katayama H and Ohgaki S. Real-time pCR detection of adenoviruses, polyomaviruses and torque viruses in river water in Japan. Water resources, 2010; 44:1747-1752

87. Fumian TM, Guimarães FR, Pereira Vaz BJ, da Silva MT, Muylaert FF, Bofill-Mas S, Gironés R, Leite JP and Miagostovich MP. Molecular detection, quantification and characterization of human polyomavirus JC from waste water in Rio De Janeiro, Brazil, 2010; 8(3):438-45

88. McQuaig S, Griffith J and Harwood VJ. The Association of Fecal Indicator Bacteria with Human Viruses and Microbial Source Tracking Markers at Coastal Beaches Impacted by Nonpoint Source Pollution Applied and Environmental Microbiology AEM.00024-12; published ahead of print 6 July 2012

Chapter 2

Optimal Irrigation Planning and Operation of Multi Objective Reservoir Using Fuzzy Logic

Jyotiba B. Gurav

Department of Civil Engineering, Amrutvahini College of Engineering, Sangamner, India

ABSTRACT

In the present study the MOFLP models have been developed for the optimal cropping pattern planning which maximizes the four objectives such as Net Benefits (NB), Crop Production (CP), Employment Generation (EG) and Manure Utilization (MU) under conflicting situation and also, for maximization of Releases for Irrigation (RI) and Releases for Power (RP) simultaneously under uncertainty by considering the fuzziness in the objective functions. The developed models have been applied using the LINGO 13 (Language for Interactive General Optimization) optimization software to the case study of the Jayakwadi Project Stage-II across Sindhphana River, in the State of Maharashtra India. The various constraints have been taken into consideration like sowing area, affinity to crop, labour availability, manure availability, water availability for optimal cropping pattern planning. Similarly constraints to find the optimal reservoir operating policy are releases for power and turbine capacity, irrigation demand, reservoir storage capacity, reservoir storage continuity. The level of satisfaction for a compromised solution of optimal cropping pattern planning for four conflicting objectives under fuzzy environment is worked out to be $\lambda = 0.68$. The MOFLP compromised solution provides NB = 1088.46 (Million Rupees), CP = 241003 (Tons), EG = 23.13 (Million Man days) and MU = 111454.70 (Tons) respectively. The compromised solution for optimal operation of multi objective reservoir yields the level of satisfaction (λ) = 0.533 for maximizing the releases for irrigation and power simultaneously by satisfying the constraint of the system under consideration. The compromised solution provides the optimal releases, i.e. RI = 348.670 Mm^3 and RP = 234.285 Mm^3 respectively.

KEYWORDS

Irrigation Planning, Reservoir Operation, Uncertainty, Multi Objective Fuzzy Linear Programming

1. INTRODUCTION

Irrigation planning objectives associated with uncertainty are conflicting and especially in developing countries like India, there are many objectives related to irrigation planning and reservoir operation that must simultaneously be satisfied. Therefore, irrigation planning with uncertainty to find optimal cropping patterns and to derive the optimal releases of reservoir operation, has thus studied in multi objective framework under fuzzy environment so that suitable and sustainable strategies can be developed for practical implementation.

The LP irrigation planning model has been developed for evaluation of irrigation development strategy for the case study of the Sri Ram Sagar project in the State of Andhra Pradesh, India. The uncertainty in inflows arising out of the uncertainty in the rainfall is tackled through chance constrained (stochastic) programming [1]. Raju and Nagesh Kumar [2] formulated the Fuzzy Linear Programming (FLP) for three conflicting objectives and dealt with fuzzification of objective functions. Raju and Duckstein [3] developed the MOFLP model for sustainable irrigation planning, considering only the objectives fuzzy in nature. Regulwar and Anand Raj [4] developed a monthly Multi Objective Genetic Algorithm Fuzzy Optimization model and it has applied to a multi- reservoir system in the Godavari river sub basin in the State of Maharashtra, India. Regulwar and Kamodkar [5] have focused on the derivation of multipurpose single reservoir release policies with fuzzy constraints. Regulwar and Gurav [6] -[10] and Gurav and Regulwar [11] [12] have presented the irrigation planning under uncertainty using multi objective fuzzy linear programming approach. Kamodkar and Regulwar [13] -[15] have presented the study of multiobjective single reservoir operation model focused with the uncertainty associated with releases for irrigation and hydropower and applied same to the Jayakwadi project stage-II in the State of Maharashtra, India.

In the literature presented, it has found that the no model has been formulated for the multi objective sustainable irrigation planning and reservoir operation with Fuzzification of the objective functions to the case study of the Jayakwadi reservoir project stage-II. The objectives of the present study are to develop the model for optimal cropping pattern planning under fuzzy environment and also to develop model for optimal operating policies in a multiple crop, multiple criterion environment, on a reservoir in a river sub basin.

2. METHODOLOGY AND MODEL DEVELOPMENT

2.1. Description of the Case Study Area

The salient features of the Jayakwadi Project Stage-II have shown in the following Table 1. Figure 1shows an index map of Jayakwadi project, State of Maharashtra, India.

2.2. Model Development for Cropping Pattern Planning

The objective is to find out the optimal cropping pattern for 75% dependable yield in command area. The problem has been formulated as an optimization model based on deterministic inflows.

2.3. Objective Function

The following four objectives for irrigation planning have been considered in the present study.

Table 1. Salient features of the Jayakwadi project stage-II.

Sr. No.	Particular	Quantity
1	Gross Capacity at F. R. L.	453.64 Mm^3
2	Capacity of Dead Storage	142.34 Mm^3
3	Capacity of Live Storage	311.30 Mm^3
4	Capacity for Power Generation	2.25 MW
5	No. of Turbines	3 × 750 KW
6	Irrigable Command Area	938.85 Km^2

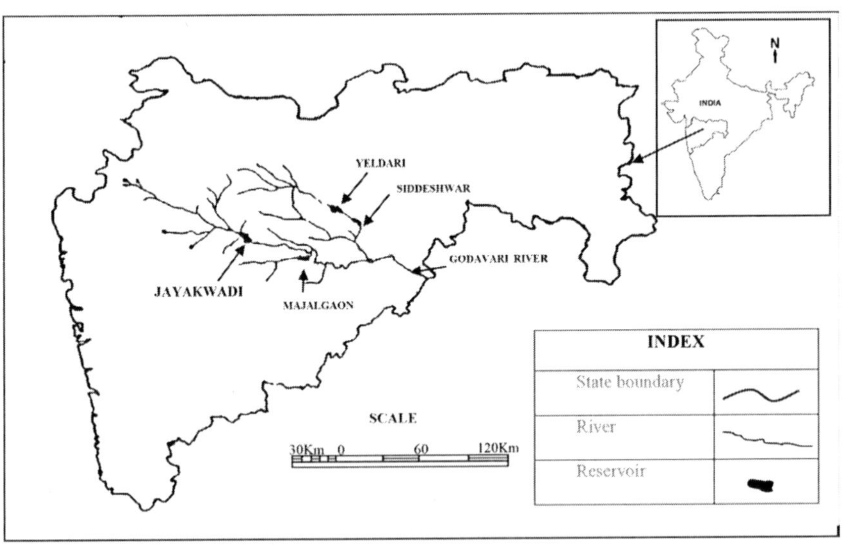

Figure 1. Index map of Jayakwadi project, state of Maharashtra, India.

2.4. Maximization of Net Benefits (NB)

Maximization of Net benefits can be expressed as:

$$\text{Maximize } NB = \left[\left(\sum_{i=1}^{K_1} A_i^K + \sum_{i=1}^{R_1} A_i^R + \sum_{i=1}^{T_1} A_i^{TS} + \sum_{i=1}^{P_1} A_i^P + \sum_{i=1}^{H_1} A_i^{HW}\right) BC_i - \left(\sum_{i=1}^{K_1} A_i^K + \sum_{i=1}^{R_1} A_i^R + \sum_{i=1}^{T_1} A_i^{TS} + \sum_{i=1}^{P_1} A_i^P + \sum_{i=1}^{H_1} A_i^{HW}\right) IC_i\right] \quad (1)$$

[In which i = crop index. 1 = Sugarcane (P), 2 = Banana (P), 3 = Chilies (TS), 4 = L S Cotton (TS), 5 = Sorghum (K), 6 = Paddy (K), 7 = Sorghum (R), 8 = Wheat (R), 9 = Gram (R) and 10 = Groundnut (HW) and ha = 10000 m^2]

A_i^K = Area of ith crop in kharif season (ha);

A_i^R = Area of ith crop in rabi season (ha);

A_i^{TS} = Area under two seasonal crops in (ha);

A_i^P = Area under perennial crop in (ha);

A_i^{HW} = Area under hot weather crop in (ha); BC_i = Benefit coefficient for i^{th} Crop;

IC_i = Input cost for i^{th} Crop; P = Perennial;

P_1 = Perennial crops; TS = Two seasonal;

T_1 = Two seasonal crops; K = Kharif;

K_1 = Crops under Kharif season; R = Rabi;

R_1 = Crops under Rabi season; HW = Hot weather;

H_1 = Hot weather crops.

The Net Benefits coefficients (NB) from the irrigated area under various crops are obtained by subtracting the input cost (20% of gross benefit) from gross benefit for different crops. The Gross benefits are calculated by multiplying the average yield of a crop per ha and current market price of that crop. The Crop Production (CP) coefficients are taken as the average yield of a crop per ha [16].

2.5. Maximization of Crop Production (CP)

The crop production is maximized and can be expressed as:

$$\text{Maximize } (CP) = \left[\left(\sum_{i=1}^{K_1} A_i^K + \sum_{i=1}^{R_1} A_i^R + \sum_{i=1}^{T_1} A_i^{TS} + \sum_{i=1}^{P_1} A_i^P + \sum_{i=1}^{H_1} A_i^{HW}\right) AY_i\right] \tag{2}$$

AY_i = Average yield of ith crop.

2.6. Maximization of Employment Generation (EG)

Keeping in mind the socioeconomic development, the policy maker has to concentrate on the maximization of employment generation or labour availability.

$$\text{Maximize } (EG) = \left[\left(\sum_{i=1}^{K_1} A_i^K + \sum_{i=1}^{R_1} A_i^R + \sum_{i=1}^{T_1} A_i^{TS} + \sum_{i=1}^{P_1} A_i^P + \sum_{i=1}^{H_1} A_i^{HW}\right) MD_i\right] \tag{3}$$

MD_i = Number of man days for i^{th} crop per ha.

The requirement of labour or numbers of Man Days (MD) for a particular crop per ha is considered through discussion with farmers and experts from agricultural fields.

2.7. Maximization of Manure Utilization (MU)

In order to maintain the fertility and nutrient sufficiency of soil in a proper manner, one should concentrate on the maximization of the use of manures.

$$\text{Maximize } (MU) = \left[\left(\sum_{i=1}^{K_1} A_i^K + \sum_{i=1}^{R_1} A_i^R + \sum_{i=1}^{T_1} A_i^{TS} + \sum_{i=1}^{P_1} A_i^P + \sum_{i=1}^{H_1} A_i^{HW}\right) MU_i\right] \quad (4)$$

MU_i = Requirement/Utilization of manure in tons for i^{th} crop per ha.

The Requirement of Manure (MU) for a crop per ha is considered through discussion with farmers and experts from agricultural fields.

2.8. Constraints

The above four objectives are subject to the various constraints like total sowing area, maximum sowing area, affinity to particular crops, labour availability, manure availability, water availability and non-negativity (The details are not given due to space limitation).

2.9. Model Development for Reservoir Operation

The two objective functions are considered for the present case study, which are:

1) Maximization of irrigation releases (i.e. RI)

$$\text{Maximize } z = \sum_{t=1}^{12} (RI)_{it} \quad (5)$$

2) Maximization of hydropower releases (i.e. RP)

$$\text{Maximize } z = \sum_{t=1}^{12} (RP)_{it} \quad (6)$$

2.10. Constraints

The above two objectives are subject to the various constraints like Releases for Power, Turbine Capacity, Irrigation Demand, Reservoir

Storage-Capacity, and Reservoir Storage Continuity Constraint (The details are not given due to space limitation).

2.11. Algorithm for Fuzzy Linear Programming with Fuzzy Objective Functions ($\tilde{Z}$)

The Fuzzy Linear Programming (FLP) Algorithm for fuzzy objectives has divided into the following steps:

1) Considering only one objective at a time and solve the problem as a linear programming problem.

2) Focusing on the results obtained in the step 1, work out the corresponding values of each objective from the solution obtained.

3) Comparing the values of objective function obtained in step 2; find out (Z_U) and (Z_L) for each objective under consideration.

4) Keeping in view the values of (Z_U) and (Z_L) for each objective; establish the linear membership function.

5) Introducing the dummy variable (λ), now the objective function changes to maximize the dummy variable (λ) subjected to the additional constraints due to the fuzziness in the value of the objective functions and original constraints.

6) Develop the equivalent Linear Programming (LP) model as a Multi Objective Fuzzy Linear Programming (MOFLP) model.

7) Work out compromised solution with the level of satisfaction (λ).

2.12. Fuzzy Linear Programming

The most general type of fuzzy linear programming is formulated as follows:

$$\max \sum_{j=1}^{n} C_j X_j$$

$$\text{S.t.} \sum_{j=1}^{n} A_{ij} X_j \le B_i \qquad (i \in \mathbb{N}_m)$$

$$X_j \ge 0 \qquad (j \in \mathbb{N}_n)$$

where A_{ij}, B_i, C_j are fuzzy numbers, and X_j are variables whose states are fuzzy numbers$(i \in \mathbb{N}_m, j \in \mathbb{N}_n)$; the operations of addition and multiplication are operations of fuzzy arithmetic, and ≤ denotes the ordering of fuzzy numbers [17] .

3. RESULTS AND DISCUSSION

In the present study the MOFLP models have been developed for the optimal cropping pattern planning which maximizes the four objectives such as NB, CP, EG and MU under conflicting situation and also, for maximization of RI and RP simultaneously under uncertainty by considering the fuzziness in the objective functions. The developed models have been applied to the case study of the Jayakwadi project stage-II across the Sindhphana River near Majalgaon town.

The objective functions of the present study considering the Equations (1)-(4) of the Linear Programming Planning (LPP) model, which are the maximization of the NB, CP, EG and MU for the command area of the Jayakwadi Project Stage-II. These objective functions are maximized separately subjected to constraints using the LINGO 13 (Language for INteractive General Optimization) optimization software. The results of this individual maximization of the four different objectives are used to construct the linear membership function for each objective taking the help of the best (+) and worst (−) value of the same. The membership functions are shown graphically in Figures 2-5.

The same membership functions are written in the form of mathematical Equations (7)-(10).

$$\mu_{z_1}(x) = \begin{cases} 1 & z_1 \ge 1180.33 \\ \dfrac{(z_1 - 892.76)}{(1180.33 - 892.76)} & 892.76 < z_1 < 1180.33 \\ 0 & z_1 \le 892.76 \end{cases} \tag{7}$$

$$\mu_{z_2}(x) = \begin{cases} 1 & z_2 \geq 315610.80 \\ \dfrac{(z_2 - 77762.07)}{(315610.80 - 77762.07)} & 77762.07 < z_2 < 315610.80 \\ 0 & z_2 \leq 77762.07 \end{cases} \tag{8}$$

$$\mu_{z_3}(x) = \begin{cases} 1 & z_3 \geq 24.86 \\ \dfrac{(z_3 - 19.43)}{(24.86 - 19.43)} & 19.43 < z_3 < 24.86 \\ 0 & z_3 \leq 19.43 \end{cases} \tag{9}$$

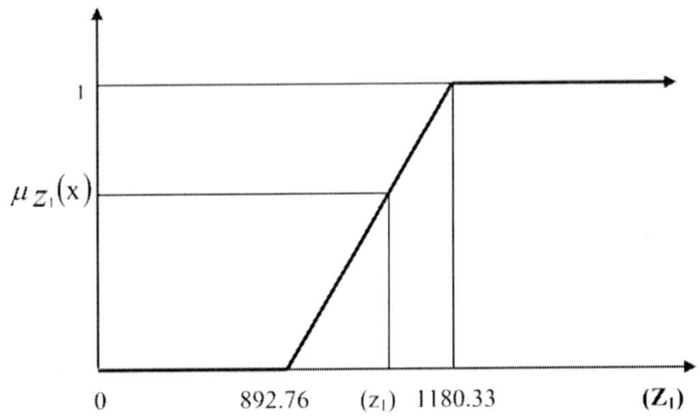

Figure 2. Membership function for Z_1 (Million rupees).

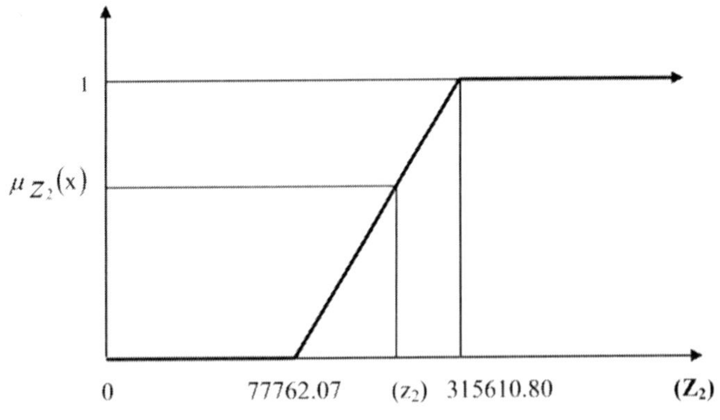

Figure 3. Membership function for Z_2 (Tons).

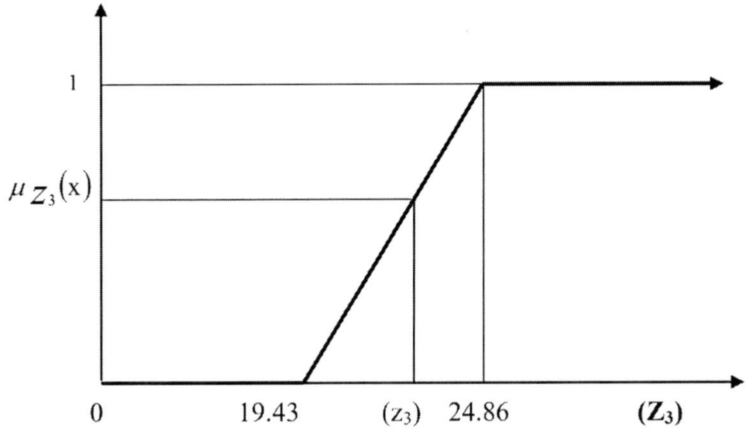

Figure 4. Membership function for Z_3 (Million man days).

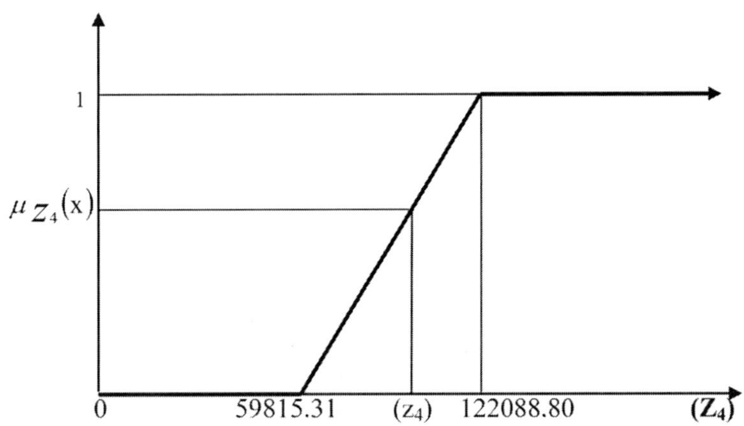

Figure 5. Membership function for Z_4 (Tons).

$$\mu_{z_4}(x)=\begin{cases}1 & z_4 \geq 122088.80\\ \dfrac{(z_4-59815.31)}{(122088.80-59815.31)} & 59815.31 < z_4 < 122088.80\\ 0 & z_4 \leq 59815.31\end{cases} \quad (10)$$

The results are shown in Table 2. The membership functions $\mu_1(X)$, $\mu_2(X)$, $\mu_3(X)$ and $\mu_4(X)$ of the fuzzy sets characterizing the objective functions rise linearly from 0 to 1 at the highest achievable value of Z_1 = 1180.33 Million Rs, Z_2 = 315610.80 Tons, Z_3 = 24.86 Million Man days and Z_4 = 122088.80 Tons respectively. The level of satisfaction associated with NB rises from 0 if the NB is 892.76 Million Rs or less to 1 if the total NB is Z_1 = 1180.33 Million Rs or more. The level of satisfaction with respect to CP rises from 0 if the CP is 77762.07 Tons or less to 1 if the CP is Z_2 = 315610.80 Tons or more and the satisfaction level associated with EG rises from 0 for 19.43 Million Man days or less to 1 for EG Z_3 = 24.86 Million Man days and more.

Similarly, the level of satisfaction associated with MU rises from 0 if the MU is 59815.31 Tons or less to 1 if the MU is Z_4 = 122088.80 Tons or more. The maximum satisfaction level of the membership functions of four participating/conflicting objectives has been designated as the 'best' achieved/compromised solution. Finally the modified form of the optimization problem (MOFLP) by introducing the dummy variable $\lambda = \min[\mu_1(X), \mu_2(X), \mu_3(X), \mu_4(X)]$ such that the objective is to:

Maximize λ

Subject to,

$$\left(Z_1 - 892.76\times10^6\right)/\left(1180.33\times10^6 - 892.76\times10^6\right) \geq \lambda$$

$$\left(Z_2 - 77762.07\right)/\left(315610.80 - 77762.07\right) \geq \lambda$$

$$\left(Z_3 - 19.43\times10^6\right)/\left(24.86\times10^6 - 19.43\times10^6\right) \geq \lambda$$

$$\left(Z_4 - 59815.31\right)/\left(122088.80 - 59815.31\right) \geq \lambda$$

and all other original constraint given in the model; $\lambda \geq 0$. The solution of MOFLP model is presented in Table 2. From this solution, it is observed that the area to be irrigated is constant in case of Chilies (TS), Sorghum (K) and Paddy (K) under LP model and MOFLP model solution. When the NB is to be maximized, then the area under irrigation is zero for LS Cotton (TS) and Groundnut (HW) because the benefit coefficient is low for these crops. Similarly, when the Crop Production is to be maximized, then the area under irrigation is zero for LS Cotton (TS) and Groundnut (HW).

Table 2. Solution for the NB, CP, EG and MU by LPP and MOFLP model.

S. N.	Crop and Season	Solution for Maximization of				Compromised Solution ($\lambda = 0.680$)
		NB (Z_1) (ha)	CP (Z_2) (ha)	EG (Z_3) (ha)	MU (Z_4) (ha)	Area of Crop (ha)
1	Sugarcane (P)	2750	2750	0	2750	1758.90
2	Banana (P)	1375	1375	0	1375	1375
3	Chilies (TS)	2750	2750	2750	2750	2750
4	L S Cotton (TS)	0	0	22895	22895	21026.67
5	Sorghum (K)	10996	10996	10996	10996	10996
6	Paddy (K)	9158	9158	9158	9157.20	9158
7	Sorghum (R)	12013.49	13737	13737	13737	0
8	Wheat (R)	22895	22895	22895	9481.37	22895
9	Gram (R)	4579	2774.70	4579	0	4579
10	Groundnut (HW)	0	0	1368.03	2750	0
	Net Cropped Area (Ha)	66516.49	66435.70	88378.03	75891.57	74538.57

NB: $Z_1 = 1180.33$ (Z_1^+) (Million Rs): $Z_2 = 315448.20$ (Tons): $Z_3 = 19.54$ (Million Man Days): $Z_4 = 59642.34$ (Tons):
Irrigation Intensity (%) = 70.85
CP: $Z_1 = 1175.63$ (Million Rs): $Z_2 = 315610.80$ (Z_2^-) (Tons): $Z_3 = 19.43$ (Z_3^-) (Million Man Days): $Z_4 = 59815.31$ (Z_4^-) (Tons):
Irrigation Intensity (%) =70.76
EG: $Z_1 = 892.76$ (Z_1^-) (Million Rs): $Z_2 = 77762.07$ (Z_2^-) (Tons): $Z_3 = 24.86$ (Z_3^+) (Million Man Days): $Z_4 = 108995.10$ (Tons):
Irrigation Intensity (%) = 94.13
MU: $Z_1 = 1059.16$ (Million Rs): $Z_2 = 300763.20$ (Tons): $Z_3 = 20.82$ (Million Man Days): $Z_4 = 122088.80$ (Z_4^+) (Tons):
Irrigation Intensity (%) = 80.83
Compromised Solution: $Z_1 = 1088.46$ (Million Rs): $Z_2 = 241003$ (Tons): $Z_3 = 23.13$ (Million Man Days): $Z_4 = 111454.70$ (Tons):
Irrigation Intensity (%) = 79.40

(+ve is the best value/aspired level of objective and −ve is the worst value/lowest acceptable level of objective)

Also, when the EG is to be maximized, then the area under irrigation is zero for Sugarcane (P), Banana (P). This is due to the labour requirement per ha is low and also due to the limited area under the existing cropping pattern. On the other hand, when MU is to be maximized, then the area under irrigation is zero for Gram (R) as manure requirement per ha is low and also due to the limited area under the existing cropping pattern. The area under irrigation is zero for Sorghum (R) and Groundnut (HW) if four conflicting objectives are considered simultaneously under MOFLP environment. In case of individual optimization for four objectives separately the irrigation intensity is 70.85%, 70.76%, 94.13% and 80.83% respectively, while in case of MOFLP it is 79.40%.

In case of MOFLP the irrigation intensity is more by 8.55%, 8.64%, 1.21% if we compare with individual optimization for net benefits and crop production respectively, and less by 14.73% and 1.43% if we compare with individual optimization for employment generation and manure utilization respectively.

The level of satisfaction for a compromised solution for four conflicting objectives under fuzzy environment is works out to be λ = 0.68. The MOFLP compromised solution provides NB = 1088.46 (Million Rupees), CP = 241003 (Tons), EG = 23.13 (Million Man days) and MU = 111454.70 (Tons) respectively.

The results of this individual maximization of the two different objectives are used to construct the linear membership function for each objective taking the help of the best (+) and worst (–) value of the same. The membership functions are shown graphically in Figure 6 and Figure 7. The same membership functions are written in the form of mathematical Equations (11-12). The results are shown in Table 3.

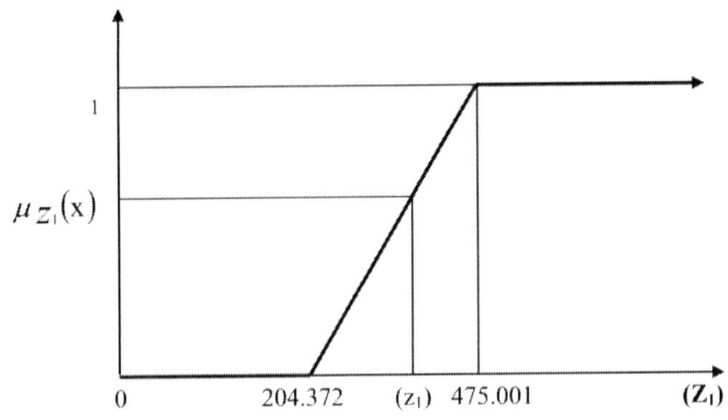

Figure 6. Membership function for Z_1 (Mm^3).

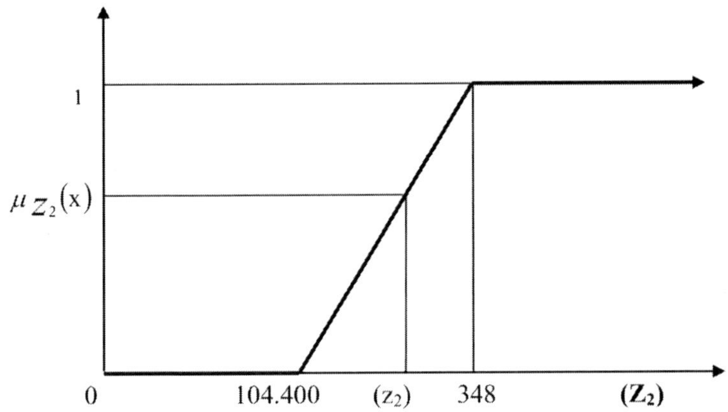

Figure 7. Membership function for Z_2 (Mm^3).

$$\mu_{z_1}(x)=\begin{cases}1 & z_1 \geq 475.001\\ \dfrac{(z_1-204.372)}{(475.001-204.372)} & 204.372<z_1<475.001\\ 0 & z_1 \leq 204.372\end{cases} \quad (11)$$

$$\mu_{z_2}(x)=\begin{cases}1 & z_2 \geq 348\\ \dfrac{(z_2-104.400)}{(348-104.400)} & 104.400<z_2<348\\ 0 & z_2 \leq 104.400\end{cases} \quad (12)$$

The membership functions $\mu_1(X)$, and $\mu_2(X)$ of the fuzzy sets characterizing the objective functions rise linearly from 0 to 1 at the highest achievable value of Z_1 = 475.001 Mm^3 and Z_2 = 348.00 Mm^3 respectively. The level of satisfaction associated with RI rises from 0 if the RI is 204.372 Mm^3or less to 1 if the total RI is Z_1 = 475.001 Mm^3 or more. The level of satisfaction with respect to RP rises from 0 if the RP is 104.400 Mm^3 or less to 1 if the RP is Z_2 = 348 Mm^3 or more. The maximum level of satisfaction from the membership functions of two participating/conflicting objectives has been designated as the 'best' achieved/compromised solution.

The other main objective functions of the present study considering the Equations (5), (6) of the LPP model, which are the maximization of the releases for irrigation and releases for hydropower of the Jayakwadi Project Stage-II. These objective functions are maximized separately subjected to constraints using the LINGO 13 (Language for INteractive General

Optimization) optimization software. Finally the modified form of the optimization problem (MOFLP) by introducing the dummy variable $\lambda = \min[\mu_1(X), \mu_2(X),]$ such that the objective is to:

Table 3. Solution for maximization of RI and RP by LPP and MOFLP model.

S. N.	Month	Solution for Maximization of Releases (Mm^3)				Compromised Solution ($\lambda = 0.533$)	
		RI	RP	RI	RP	RI (Mm^3)	RP (Mm^3)
1	June	2.929	8.700	2.136	29.000	2.136	9.493
2	July	20.830	8.700	6.249	29.000	6.249	27.960
3	August	37.640	8.700	11.292	29.000	37.640	29.000
4	September	46.020	8.700	13.806	29.000	46.020	29.000
5	October	132.010	8.700	39.603	29.000	125.339	29.000
6	November	127.050	8.700	38.115	29.000	38.115	29.000
7	December	42.180	8.700	26.829	29.000	26.829	29.000
8	January	30.204	8.700	30.204	29.000	30.024	17.032
9	February	9.006	8.700	9.006	29.000	9.006	8.700
10	March	8.694	8.700	8.694	29.000	8.694	8.700
11	April	10.674	8.700	10.674	29.000	10.674	8.700
12	May	7.764	8.700	7.764	29.000	7.764	8.700
	Total	475.001 (Z_{1Max})	104.400 (Z_{2Min})	204.372 (Z_{1Min})	348.000 (Z_{2Max})	348.670	234.285

Maximize λ

Subject to

$$\left(Z_1 - 204.372\times10^6\right)/\left(475.001\times10^6 - 204.372\times10^6\right) \geq \lambda$$

$$\left(Z_2 - 104.400\times10^6\right)/\left(348\times10^6 - 104.400\times10^6\right) \geq \lambda$$

and all other original constraint in the model; λ ≥ 0. The solution of MOFLP is presented in Table 3. The compromised solution yields the level of satisfaction (λ) = 0.533 for maximizing the releases for irrigation and power simultaneously by satisfying the constraint of the system under consideration. The compromised solution provides the optimal releases, i.e. RI = 348.670 Mm^3 and RP = 234.285 Mm^3 respectively.

4. CONCLUSIONS

The objective of the study is to develop the MOFLP models for optimal cropping pattern that maximizes Net Benefits, Crop Production, Employment Generation and Manure Utilization simultaneously and also for maximization of releases for irrigation and power for optimal operation of the reservoir. To achieve this, max-min approach based MOFLP models have been developed and applied to Jayakwadi Project Stage II, Maharashtra State, India.

The observations from the present study are as given below:

Ÿ This study proposes a basis for irrigation planning as an integrated approach.

Ÿ The level of satisfaction for a compromised solution for four conflicting objectives under fuzzy environment is works out to be λ = 0.680. The MOFLP compromised solution provides NB = 1088.46 Million Rupees, CP = 241003 Tons, EG = 23.13 Million Man days and MU = 111454.70 Tons respectively and irrigation intensity is 79.40%.

Ÿ The results of the application of MOFLP model for reservoir operation indicate that the maximum level of satisfaction (λ) is 0.533 achieved by maximizing both the objectives simultaneously and the corresponding values of RI and RP are 348.670 Mm^3 and 234.285 Mm^3 respectively.

The present model will be helpful for the decision maker to take decisions under conflicting situations when planning for different objectives simultaneously.

Ÿ The developed models are capable to trace out an integrated irrigation planning with prime consideration for economic, social and environmental issues.

Ÿ The results obtained under the present study are sensitive to the changes in the market price of the crop, cost of crop production, unit labour cost, and unit manure cost and water availability.

ACKNOWLEDGEMENTS

The author is thankful to the Director, BCUD of Savitribai Phule Pune University, Pune State of Maharashtra, India for permission to take up this study at Amrutvahini College of Engineering, Sangamner and for providing all the necessary facilities and research grant for carrying out the research work. The author would like to express sincere thanks to Command Area Development Authority, Aurangabad, Maharashtra State, India and Mahatma Phule Krishi Vidyapeeth Rahuri, Ahmednagar, Maharashtra State, India for providing necessary data for analysis.

REFERENCES

1. Raju, K.S. and Nagesh Kumar, D. (2000) Optimum Cropping Pattern for Sri Ram Sagar Project: A Linear Programming Approach. Journal of Applied Hydrology, XIII, 57-67.

2. Raju, K.S. and Nagesh Kumar, D. (2000) Irrigation Planning of Sri Ram Sagar Project Using Multi Objective Fuzzy Linear Programming. ISH Journal of Hydraulic Engineering, 6, 55-63.

3. Raju, K.S. and Duckstein, L. (2003) Multiobjective Fuzzy Linear Programming for Sustainable Irrigation Planning: An Indian Case Study. Soft Computing, 7, 412-418.

4. Regulwar, D.G. and Anand Raj, P. (2008) Development of 3-D Optimal Surface for Operation Policies of a Multireservoir in Fuzzy Environment Using Genetic Algorithm for River Basin Development and Management. Water Resources Management, 22, 595-610.

5. Regulwar, D.G. and Kamodkar, R.U. (2010) Derivation of Multipurpose Single Reservoir Release Policies with Fuzzy Constraints. Journal of Water Resource and Protection, 2, 1030-1041.

6. Regulwar, D.G. and Gurav, J.B. (2010) Fuzzy Approach Based Management Model for Irrigation Planning. Journal of Water Resource and Protection, 2, 545-554.

7. Regulwar, D.G. and Gurav, J.B. (2011) Irrigation Planning under Uncertainty—A Multi Objective Fuzzy Linear Programming Approach. Water Resources Management, 25, 1387-1416.

8. Regulwar, D.G. and Gurav, J.B. (2012) Sustainable Irrigation Planning with Imprecise Parameters under Fuzzy Environment. Water Resources Management, 26, 3871-3892.

9. Regulwar, D.G. and Gurav, J.B. (2013) Two-Phase Multiobjective Fuzzy Linear Programming Approach for Sustainable Irrigation Planning. Journal of Water Resource and Protection, 5, 642-651.

10. Regulwar, D.G. and Gurav, J.B. (2014) Irrigation Planning with Fuzzy Parameters—An Interactive Approach. Journal of Agricultural Science and Technology, 16, 1157-1172.

11. Gurav, J.B. and Regulwar, D.G. (2011) Minimization of Cost of Cultivation along with Interactive Decision Making under Fuzzy Environment. ISH Journal of Hydraulic Engineering, 17, 30-42.

12. Gurav, J.B. and Regulwar, D.G. (2012) Multi Objective Sustainable Irrigation Planning with Decision Parameters and Decision Variables Fuzzy in Nature. Water Resources Management, 26, 3005-3021.

13. Kamodkar, R.U. and Regulwar, D.G. (2010) Modelling Multiobjective Reservoir Operation in Uncertain Parameter Environment. Journal of Indian Water Works Association, 8, 27-35.

14. Kamodkar, R.U. and Regulwar, D.G. (2010) Multipurpose Single Reservoir Operation with Fuzzy Technological Constraints. ISH Journal of Hydraulic Engineering, 16, 49-63.

15. Kamodkar, R.U. and Regulwar, D.G. (2014) Optimal Multiobjective Reservoir Operation with Fuzzy Decision Variables and Resources: A Compromise Approach. Journal of Hydro-Environment Research, 8, 428-440.

16. The Report of Commissionerate of Agriculture, Maharashtra State (2006) Agricultural Statistical Information Part-II, Maharashtra State, India, Part-II: 1-255.

17. Klir, G.J. and Yuan, B. (2007) Fuzzy Sets and Fuzzy Logic: Theory and Applications. Prentice Hall of India Private Ltd, New Delhi.

Chapter 3

Hydrogeochemical Assessment of Groundwater in the Precambrian Rocks, South Eastern Desert, Egypt

Ashraf Embaby[1,2], Moumtaz Razack[1*], Mathieu Lecoz[1], Gilles Porel[1]

[1]Department of Hydrogeology, University of Poitiers, Poitiers, France

[2]Geology Department, Faculty of Science, Sohag University, Sohag, Egypt

ABSTRACT

This work refers to the characterization of the hydrochemistry of the southern part of the Eastern Desert in Egypt, on the basis of physico-chemical properties of groundwater occurring in the fractured Precambrian rocks inland and in sedimentary formations on the coastline of the Red Sea. Thirty-five groundwater samples have been collected from the study area for hydrochemical investigations to understand the sources of dissolved ions and assess the chemical quality of the groundwater. Several methods were used to interpret the hydrochemical data, i.e. graphical methods, principal components analysis, ions exchanges indices and saturation indices of various minerals. The results show that the major ionic relationships are $Na^+ > Ca^{2+} > Mg^{2+}$ and $Cl^- > SO_4^{2-} > HCO^{3-}$ and that groundwater chemical characteristics are controlled by natural geochemical processes but also, to a lesser extent, by anthropogenic activities. Natural minerals dissolution, ion exchanges and evaporation play a prominent role in the ion enrichment of groundwater. A comparison of groundwater quality in relation to WHO water quality standards proved that most of the water samples are not totally suitable for drinking water purpose.

KEYWORDS

Eastern Desert, Pre-Cambrian Rocks, Groundwater, Hydrochemistry, Principal Components Analysis, Egypt

1. INTRODUCTION

Groundwater resource assessments and sustainability considerations are of utmost importance in the arid and semiarid regions, where water is commonly of critical economical and social significance. Egypt is facing this situation. In this country water needs are dramatically increasing, demanded by a rapidly growing population, increased urbanization, higher standards of living and agricultural policy. The World Bank describes Egypt as a country being under water stress. Limited natural fresh water resources in the Eastern Desert of Egypt call for evaluation of water resources in the area. The Eastern Desert represents 22% of the total area of the country. It houses both residential areas, commercial and industrial zones, as well as tourist villages. Groundwater is a significant and crucial resource in many areas in Egypt and it commonly plays a key role as a water supply for drinking, irrigation and industries. It represents a vital resource in the Eastern Desert due to the unavailability of other sources.

Despite importance of groundwater in the study area, little is known about the natural phenomena that govern the chemical composition of groundwater or anthropogenic factors that presently affect them. The chemical composition of groundwater is controlled by many factors that include composition of precipitation, geological structure and mineralogy of the watersheds and aquifers, and geochemical processes within the aquifer [1] . The interaction of all factors leads to various water types. Thus the purpose of the present study is to determine the current hydrochemical status of the southeastern desert groundwater and understand the mineralization processes. Major chemical ions (Ca^{2+}, Mg^{2+}, Na^{+}, K^{+}, Cl^{-}, HCO^{3-}, NO^{3-}, SO_4^{2-}) were analyzed for this purpose. Samples were collected in 35 wells distributed all over the study area. The interpretation of hydrochemical data was made using various methods (Piper diagram, principal component analysis, base exchanges index, bi-ele- ments scatter diagrams, saturation indices, and mapping). The objective aims at improving the knowledge of geochemical evolution of groundwater in this arid and semi-arid region, leading to sustainable development and effective management of groundwater resource.

2. LOCATION OF THE STUDY AREA

The Eastern Desert covers about 220,000 km^2 and is located between the Red Sea and the Nile Valley. It extends from latitude 22 N and 29 N or Egypt's borders with Sudan in the South, to the Delta in the North [2] . It forms around 22% of Egypt's surface area, and extends on average over 1100 km from North to South, and 225 km from East (Nile Valley) to West (Red Sea). But it was undeveloped due to the limited availability of water [3] . The Eastern Desert consists of high rugged mountains parallel to the coast and flanked to the north and west by dissected sedimentary plateau [4] . The study area (Figure 1) is located in the Southern part of the Eastern Desert (South Eastern Desert) and extends from El-Baramia area in Idfu-Marsa Alam road to Marsa Alam-Abu Ghuson on the Red Sea coast. The study area covers an area about 16800 km^2 and is limited by longitudes 33˚50/ - 35˚45/E and latitudes 24˚00/ - 25˚15/N.

The Eastern Desert is located in the extremely arid provinces in Egypt. Temperature ranges from 21˚C to 41˚C. Rainfall is considerably variable from one season to another. The average annual rainfall ranges from 3 to 50 mm/year on the southern part of the desert.

3. GEOLOGICAL AND HYDROGEOLOGICAL SETTING

The Eastern Desert is geologically composed mainly of sedimentary and Pre-Cambrian crystalline rocks. The crystalline rocks of the Pre-Cambrian basement complex form massive formations extending parallel to the Red Sea graben and consist mainly of metamorphic and igneous rocks [4] [5] . The sedimentary rocks are found on the coastal plain and overlain the Pre-Cambrian basement rocks. They consist of Quaternary alluvial sediments and Tertiary formations (sandstones interbedded with clays, evaporites intercalated with clay and sand) [6] [7] . The quaternary aquifer in the Eastern Desert is represented by wadi sediments and alluvial plains [2] .

The groundwater occurrence in fractured igneous and metamorphic water bearing formations in the Eastern Desert of Egypt is controlled by the presence, width and direction of the joints, fractures, faults and dykes and thickness of weathering zones [8] [9] . The hydrogeological rock units in South Eastern Desert are classified into Red Sea Coastal aquifers formed by coastal Quaternary and Miocene formations, and fractured Pre-Cambrian basement aquifer composed of meta-volcanics, basic-ultrabasic rocks, granites and metasediments (schists and gneisses) (Table 1 and Figure 1). Embaby et al. [10] using geophysical methods, concluded that the structure of the study area can be schematized with three main layers, i.e. the wadi deposits 15 - 60 m thick, the fractured basement rock 20 - 100 m thick and massive basement rocks 40 - 250 m thick.

At the scale of the Eastern Desert, fractured Precambrian formations can be considered as a continuous medium.

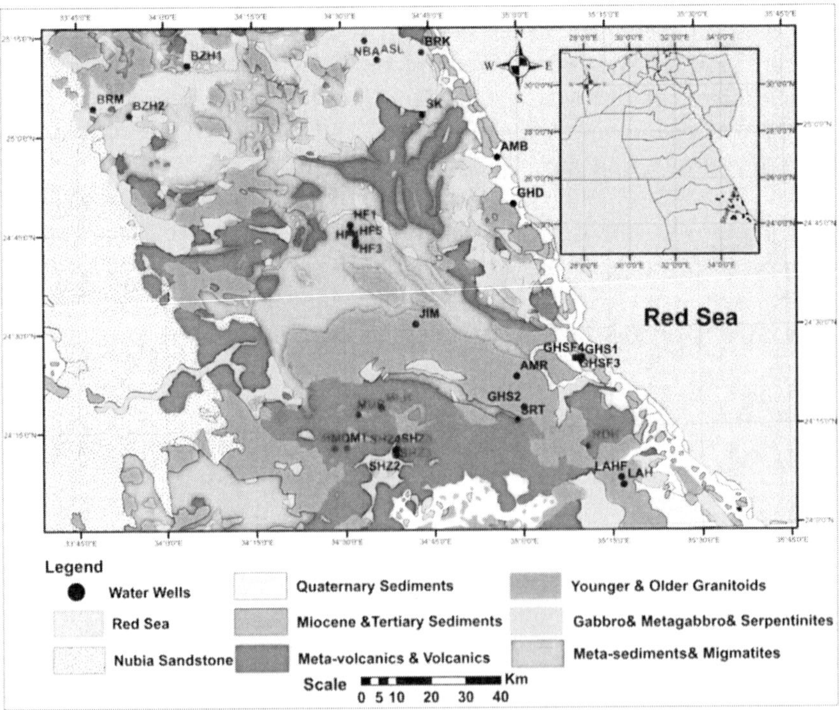

Figure 1. Simplified geological map of the study area and groundwater samples locations (modified from Aswan geological map, 1979).

A piezometric map (Figure 2) throughout the study area has thereby been established. It highlights a groundwater divide line from South to the North. To the west of this line, groundwater flows to the Nile. To the east the flow is towards the Red Sea. Recharge takes place from the occasional rainfalls. The mountainous zones probably act as preferential recharge zones.

4. SAMPLING AND ANALYTICAL PROCEDURE

A sampling network was set up to allow collection of representative data of the variability in space of the groundwater quality. This network consisted of 35 wells and covers the whole area (Figure 1). The rocks types of the aquifers where the wells were drilled are shown in Table 1.

Water samples were collected during two field campaigns, June 2013 and March 2014 in wells that form the observation network, because of great difficulties to access to the sites. Groundwater samples were filtered and collected in 0.5 L polyethylene bottles. All samples were stored in ice-boxes and then in a refrigerator in Egypt, before being sent to Poitiers in France. The analyses focused on the most common and most abundant ions in groundwater. The analyses were performed at the University of Poitiers for alkalinity and Cl^- using titration methods, and at the University of Pau, using an ICP-MS (Inductively Coupled Plasma Mass Spectrometry) for other cations and anions (Ca^{2+}, Mg^{2+}, Na^+, K^+, HCO_3^-, NO_3^-, SO_4^{2-}, Br^-). Note that for dosing $Cl^-$by titration with $AgNO_3$, to avoid interference with sulfates, one to two drops of hydrogen peroxide (H_2O_2) at 30% were added while maintaining agitation. Temperature and pH were measured in the field. The ion-balance error computation, taking the relationship between the total cations (Ca^{2+}, Mg^{2+}, Na^+, K^+) and the total anions (NO_3^-, SO_4^{2-}, HCO_3^- and Cl^-) for each water sample, is observed to be within the range of acceptability (±5%) used in most laboratories [11] , for all water samples. The hydrochemical data are given inTable 2.

5. MAJOR IONS CHEMISTRY

Table 3 shows the summary statistics for each water quality parameter. The pH ranges between 7.2 and 8.3. The average pH is 7.7, which indicates that the eastern desert groundwater is slightly alkaline. The lower pH values are located at the eastern part of the study area and reflect the connection between marine water and the sampled wells. The higher pH values of the western part of the study area may indicate preferential recharge areas. The groundwater temperature does not vary much and is included between 28°C and 29°C. The TDS (Total Dissolved Solids) ranges from 947 mg/l to 13,504 mg/l with a mean of 5974 mg/l. These figures show that the groundwater mineralization varies greatly on the study site.

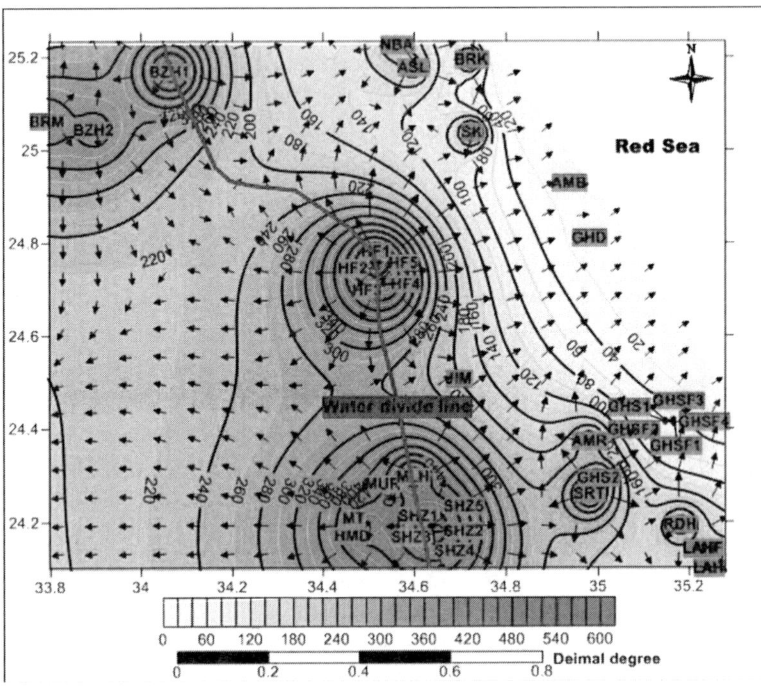

Figure 2. Piezometric map of the study area.

The spatial variability of the ions contents, expressed using the coefficient of variation (CV), is significantly high (60% < CV < 103%). Chloride concentrations range from 270 mg/l to 5378 mg/l, bromide from 0 to 30 mg/l, sulfate from 197 to 4676 mg/l, bicarbonate from 45 to 1345 mg/l, calcium from 27 to 1550 mg/l, magnesium from 28 to 1062 mg/l, potassium from 3 to 102 mg/l and sodium from 171 to 3582 mg/l. The relative concentrations of the cations occur in the order of Na^{+}, Ca^{2+}, Mg^{2+}, K^{+} and of the anions in the order of Cl^{-}, SO_4^{2-}, HCO_3^{-}. No nitrate was found in the groundwater samples.

Table 1. Aquifer rock types and groundwater samples.

No.	Aquifer rock units	Wells No.	Wells locations
1	Meta-volcanic (Meta-andesite, meta-basalt)	1, 4, 5, 6, 7, 8, 9, 10, 11, 12 and 26	Sukkari, El-Shazly$_{1-5}$, Abu Himimid, Miteiwit, Murirah, Malahaia and El-Barramia
2	Basic-ultrabasic (Gabbro, serpentinites, Meta-gabbro-diorite complex)	25, 26x and 30	Abu Ghuson-2, Sartut and El-Nabaa
3	Granitoids (Younger and older granite rocks)	13, 14, 15, 24, 27, 28, 29, 31, 32, 33, 34 and 35	El-Jimal, Beizah$_{1-2}$, El-Sheikha Amira, Hafafit$_{3-5}$, El-Asly, Lahmi, Lahmi-FAO, El-Reidah and Barikah
4	Meta-sediment (Schists, gneisses)	2 and 3	Hafaifit$_{1-2}$
5	Quaternary (Alluvium, recent sediments)	16, 17, 19, 20, 21 and 22	Ambaout, Ghadir, Abu Ghuson FAO$_{1-4}$
6	Miocene (Coral reefs, sandstones, carbonate, evaporites)	18	Abu Ghuson-1

Table 2. Hydrochemical data of the South Eastern Desert groundwater samples. TDS and ions concentrations are in mg/l. Hardness is in French degree (˚F).

Sample	TDS	Ca^{2+}	Mg^{2+}	Na^{+}	K^{+}	HCO_3^-	Cl^-	SO_4^{2-}	Br^-	pH	Hardness
SK	2656.0	188.1	94.0	613.0	7.0	414.8	812.4	763.4	0.8	7.7	86
HF1	2880.0	27.3	56.0	872.0	29.0	503.3	627.7	949.1	1.0	7.9	30
HF2	3104.0	73.5	80.0	977.0	24.0	553.9	664.7	1149.7	1.1	7.5	51
SHZ1	5017.6	547.7	104.0	1248.0	8.0	244.0	1415.5	2122.7	3.7	7.8	179
SHZ2	10304.0	699.3	115.0	2977.0	47.0	230.0	3631.1	3374.1	10.6	7.7	222
SHZ3	6848.0	657.6	135.0	1800.0	13.0	161.7	1883.2	3122.7	4.9	7.5	220
SHZ4	4115.2	511.2	91.0	996.0	3.0	137.3	1132.4	1922.1	3.4	7.6	165
SHZ5	12864.0	1009.3	77.0	3582.0	36.0	101.9	5194.3	3248.4	14.7	7.7	284
HMD	2592.0	194.9	38.0	727.0	5.0	274.5	555.8	1122.7	2.5	7.9	64
MT	3532.8	261.9	74.0	982.0	5.0	291.6	997.0	1257.4	3.1	7.6	96
MUR	3328.0	252.3	78.0	880.0	6.0	268.4	726.2	1556.8	1.7	7.6	95
MLH	3648.0	541.8	131.0	653.0	6.0	259.3	1083.2	1491.0	1.2	7.3	189
JIM	947.2	96.7	31.5	171.9	4.5	158.6	270.8	197.6	0.4	7.8	37
BZH1	13504.0	825.6	1062.0	2879.0	76.0	109.8	5378.9	4676.5	5.6	7.8	643
BZH2	9408.0	859.4	402.0	2314.0	34.0	536.8	2954.1	4089.7	12.5	7.2	380
AMB	7872.0	1014.9	426.0	1252.0	57.0	96.4	3200.3	2248.4	8.3	7.8	428
GHD	7616.0	906.4	336.0	1044.0	39.0	78.1	3483.4	754.5	22.3	7.5	364
GHS1	10944.0	1550.3	343.0	1834.0	27.0	123.2	5366.6	1284.4	30.9	7.5	528
GHSF1	5331.2	609.2	180.0	1106.0	12.0	103.7	2412.5	838.3	7.9	7.7	226
GHSF2	9216.0	920.6	304.0	1814.0	32.0	125.1	4480.4	805.4	7.4	7.5	355
GHSF3	5504.0	615.5	181.0	1112.0	14.0	311.1	2486.4	772.4	8.4	7.9	228

GHSF4	10368.0	1136.0	344.0	2071.0	13.0	97.6	4665.0	1769.4	13.0	7.4	425
AMR	1881.6	130.2	30.2	427.9	7.7	225.7	637.5	296.4	0.6	8.0	45
GHS2	3788.8	461.6	177.0	689.0	13.0	413.6	1329.3	1065.8	1.2	7.6	188
SRT	5171.2	1014.3	28.0	967.0	41.0	62.8	1944.8	1829.3	3.9	8.1	265
BRM	3219.2	186.7	107.0	927.0	24.0	707.6	366.0	1892.2	0.8	8.3	91
HF3	2540.8	121.7	81.0	706.0	16.0	329.4	430.8	1248.5	0.7	7.8	64
HF4	3180.8	74.9	101.0	1012.0	39.0	359.9	787.8	991.0	1.0	8.1	60
HF5	8448.0	235.9	221.0	2778.0	102.0	1345.4	2117.1	3517.9	3.4	8.3	150
NBA	5056.0	909.1	102.0	702.0	28.0	128.0	2075.9	1173.6	9.8	8.1	269
ASL	6240.0	903.0	88.0	1172.0	35.0	45.8	2999.5	637.7	20.1	7.7	262
LAH	2969.6	242.8	75.0	691.0	7.0	155.6	1193.9	476.0	2.8	7.6	91
LAHF	10880.0	926.9	407.0	2000.0	53.0	768.6	5120.4	724.5	17.1	7.4	399
RDH	4672.0	907.0	127.0	740.0	7.0	125.1	1749.0	1502.9	4.2	7.3	279
BRK	9472.0	1354.5	392.0	1312.0	34.0	79.3	4887.5	670.6	21.3	7.2	499

Table 3. Summary statistics of groundwater physical and chemical parameters. Ion concentrations and TDS are in mg/l. SD: Standard Deviation; CV: Coefficient of Variation.

	Ca^{2+}	Mg^{2+}	Na^{+}	K^{+}	HCO_3^-	Cl^-	SO_4^{2-}	Br^-	TDS	T°C	pH
Max	1550.3	1062.0	3582.0	102.0	1345.4	5378.9	4676.5	30.85	13504.0	29.0	8.3
Min	27.3	28.0	171.9	3.0	45.7	270.7	197.6	0.38	947.2	28.0	7.2
Average	599.1	189.1	1315.1	25.8	283.6	2258.9	1586.9	7.21	5974.8	28.4	7.7
SD	400.1	192.8	788.7	21.8	256.4	1646.8	1086.7	7.46	3332.1	0.49	0.2
CV (%)	66.8	101.9	59.9	84.5	90.4	72.91	68.48	103.4	55.7	1.7	3.6

Pearson's correlation matrix [12] was used to find relationships between the elements pairwise. The correlation matrix is shown in Table 4. TDS is strongly correlated with calcium (R = 0.77), magnesium (R = 0.71), sodium (R = 0.89), chloride (R = 0.95), sulfate (R = 0.58) and bromide (R = 0.69). These relationships clearly identify the main elements contributing to the groundwater salinity and their tendency to follow a similar trend (e.g. due to concentration by evaporation). These elements concentration tend to increase as the salinity of the groundwater increases. The salinization of the groundwater would be expected to result from the ionic concentrations increasing due as well to evaporation of recharge water, to seawater intrusion as to the effects of interactions between the groundwater and the geological formations. The strongest correlations between elements of opposite sign combine Cl^- and Na^+ (R = 0.73), Cl^- and Mg^{2+} (R = 0.69), and Cl^- and Ca^{2+} (R = 0.85). The Na^+ vs. Cl^- relationship suggests dissolution of halite or seawater intrusion. The dissolution of halite in water releases equal concentrations of sodium and chloride into the solution:

$$NaCl \rightarrow Na^+ + Cl^- \quad (1)$$

The strong relationships Cl^--Mg^{2+} and Cl^--Ca^{2+} suggest that cation exchange can also significantly affect groundwater composition. The positive and significant correlations between sulfate and sodium (R = 0.76) may indicate the contributions of evaporitic salts. Agricultural activity may also contribute to these elements. Fertilizers based on potassium sulfate, ammonium sulfate and the sulfo-phosphate ammonium [13] may contribute to these elements. Other associations between ions are further interpreted in terms of processes and origin of the mineralization of the groundwater and its evolution.

6. PRINCIPAL COMPONENTS ANALYSIS

Multivariate methods have been widely used to study the sources of variation of groundwater chemistry. Principal Components Analysis (PCA) is a multivariate statistical method, which helps to reduce a large data set with underlying linear structures into a set of variables, called Principal Components, without sacrificing much of the original information. PCA proved useful for discovering previously unsuspected relationships in the original data.

In this study a Principal Components Analysis (PCA) was performed on the chemical data for better understanding the groundwater hydrochemistry. Multivariate techniques have been applied to groundwater hydrochemistry by several authors and proved quite efficient to understand a number of geochemical processes [14] - [22] .

The extraction method, i.e. "Principal Component method" looks for a solution that maximizes the explained variance with orthogonal components, which are independent of each other. The methods to help to choose the number of components are based on relations between the eigenvalues. According to the Kaiser criteria, eigenvalues larger than one [23] , explaining more variance than the average component, should be kept. An additional graphical method can also be used, the Scree diagram (or the elbow criterion). In this diagram, the eigenvalues are plotted vs. the number of the components. If the points on the graph tend to level out (show an "elbow"), these eigenvalues are usually close enough to zero that they can be ignored. The data consist of 9 variables

(Ca^{2+}, Mg^{2+}, Na^{+}, K^{+}, Cl^{-}, SO_4^{2-}, HCO_3^{-} and TDS and pH) analyzed on all 35 samples collected throughout the study area. Eigen values, percentage of the variance of each principal component (PC) and cumulative percentage of variance of five first PCs are given in Table 5. The first two eigenvalues are >1. The scree plot is shown in Figure 3. This figure indicates, according to the elbow and Kaiser criteria, that the first two PCs should be kept for further consideration. They account for 74.7% of the total variance and are assumed to provide an adequate representation of the overall variance of the data set. Hence, in the factor matrix only these two factors are considered.

The 1st PC (Principal Component) accounts for 52.0% of the variance in the data set. PC1 is interpreted as relating mainly to the mineralization of the groundwater as it is associated to Cl^{-}, Ca^{2+}, Mg^{2+}, Na^{+}, K^{+} and TDS. Loadings for these elements on PC1 are high (Table 6). PC1 therefore opposes highly mineralized samples to weakly mineralized samples. PC2 accounts for 22.7% of the data variance. It is mainly associated to HCO_3^{-} and pH. These elements have the highest loadings on this factor. PC2 is thus related to alkaline waters. The plane associated with PC1 and PC2 accounts for 74.7% of the total variance and is accordingly quite representative of the initial data variability. Figure 4 and Figure 5 show the plot of the variables and of the samples on this plane. PC1-PC2 plane discriminates several groups of samples. It discriminates less mineralized waters lying inland in granitic, gneissic and schistic aquifers (Hafafit, Wadi-el-Jimal, Amira, ...), highly mineralized waters located close to the sea (Abu Ghuson, ...), alkaline waters marked in bicarbonates (El Barramia, Beizah, Hafafit, ...), and waters with lower pH marked in calcium and bromide (El Asly, Barika, Ghadir, Abu Ghuson, ...).

Table 4. Correlation matrix between chemical variables. Significant correlation coefficients are in bold.

	TDS	Ca^{2+}	Mg^{2+}	Na^+	K^+	Cl^-	Br^-	SO_4^{2-}	HCO_3^-
TDs	1	**0.77**	**0.71**	**0.89**	**0.62**	**0.95**	**0.69**	**0.58**	−0.07
Ca^{2+}		1	0.49	0.45	0.27	**0.85**	**0.82**	0.21	**−0.44**
Mg^{2+}			1	0.5	0.56	**0.69**	0.38	0.47	−0.05
Na^+				1	0.65	**0.73**	0.42	**0.76**	0.17
K^-					1	0.51	0.28	0.52	0.42
Cl^-						1	**0.8**	0.32	−0.23
Br^-							1	0.01	−0.26
SO_4^{2-}								1	0.19
HCO_3^-									1

The pH of most natural waters is controlled by reactions involving the carbonate system. The relation on PC2 between pH and carbonates suggest that the high pH groundwaters could be the result of calcite dissolution or precipitation. Dissolution of a small amount of calcite from minor cements may overcome the buffering capacity of fresh groundwater allowing its pH to rise. The dissolution of the calcite mineral can take place as:

$$CaCO_3 + CO_2 + H_2O \rightarrow Ca^{2+} + 2HCO_3^- \quad (2)$$

7. SPATIAL DISTRIBUTION OF TDS AND OF THE MAIN ELEMENTS

Figure 6 shows the spatial distribution of the main elements in the groundwater. The spatial distribution of concentrations depends on several factors such as lithology, the hydrodynamics of the water, the depth of the water table, climate conditions and anthropogenic pollution sources. Figure 6 shows that Cl^-, Ca^{2+} and Na^+ display relatively equivalent spatial pattern. The highest concentrations of these elements are located along the coastline and on the high peaks of the Red Sea mountains. Low contents are located at the center of the study area, where mainly meta-sediments are found.

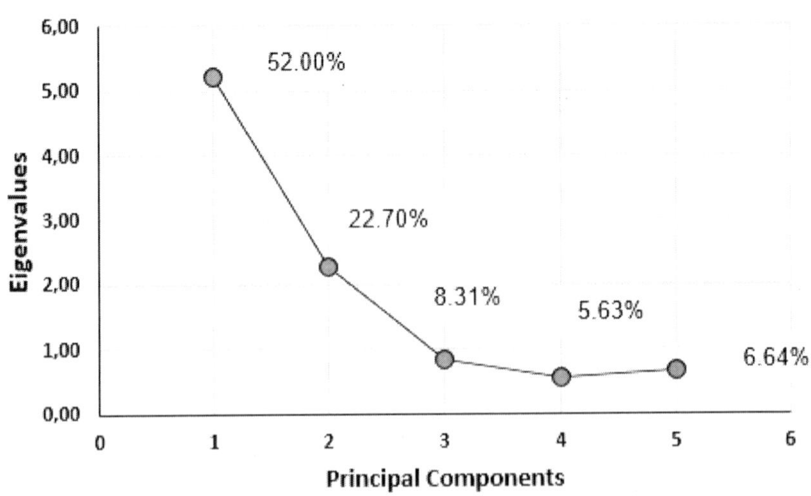

Figure 3. Principal components analysis scree plot.

Table 5. Principal component analysis, eigenvalues and explained variance.

Principal component	Eigen values	% Total variance	Cumulative eigenvalues	Cumulative %
1	5.200054	52.00	5.200054	52.00
2	2.270601	22.70	7.470655	74.70
3	0.830391	8.30	8.301045	83.01
4	0.563402	5.63	8.864447	88.64
5	0.664378	6.64	9.528826	95.28

Table 6. Chemical variables loadings on principal components 1 and 2.

Variables	PC1	PC2
TDS	**0.989**	0.054
Ca^{2+}	**0.813**	−0.449
Mg^{2+}	**0.755**	0.105
Na^{+}	**0.832**	0.376
K^{+}	0.632	**0.630**
HCO_3^-	−0.109	**0.774**
Cl^-	**0.956**	−0.184
SO_4^{2-}	0.561	0.553
Br^-	**0.739**	−0.428
pH	−0.338	**0.628**

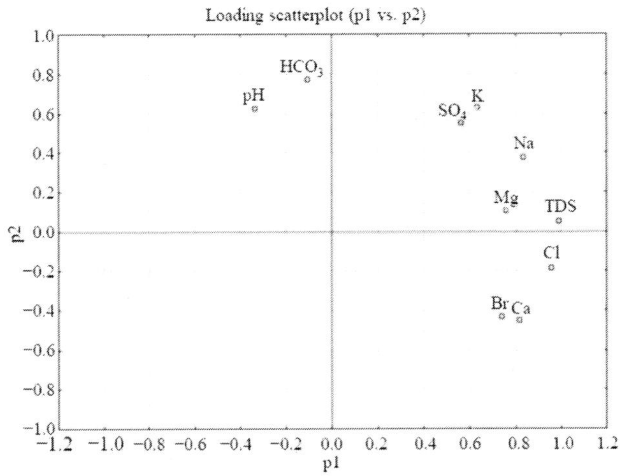

Figure 4. Principal components analysis. Plot of the variables on PC1 vs. PC2.

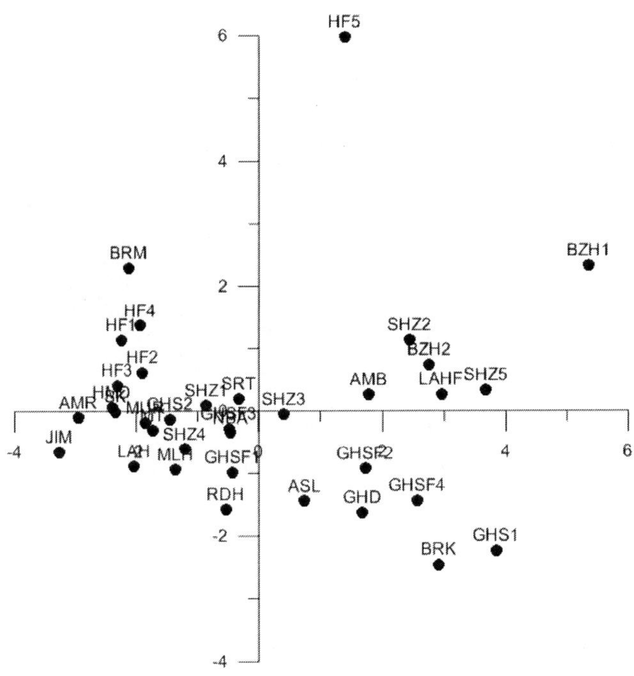

Figure 5. Principal components analysis. Plot of the groundwater samples on PC1 vs. PC2.

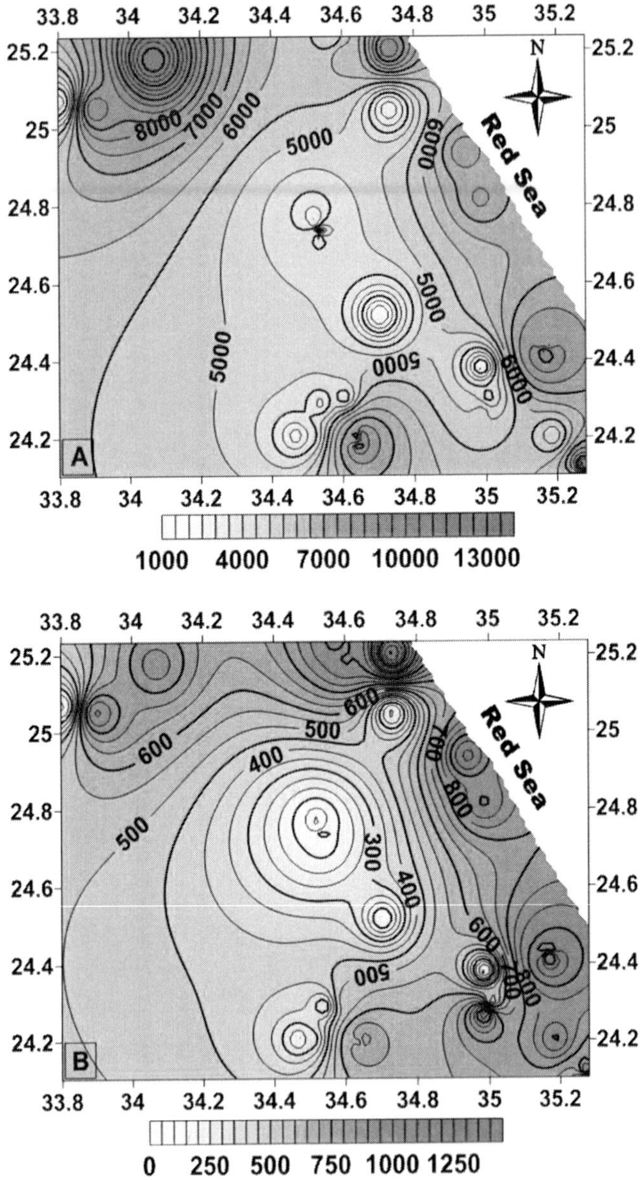
33.8
34
34.2
34.4
34.6
34.8
35
35.2
25.2
25
24.8
24.6
24.4
24.2
N
Red Sea
8000
7000
6000
5000
A
1000 4000 7000 10000 13000
600
500
400
300
700
800
B
0 250 500 750 1000 1250

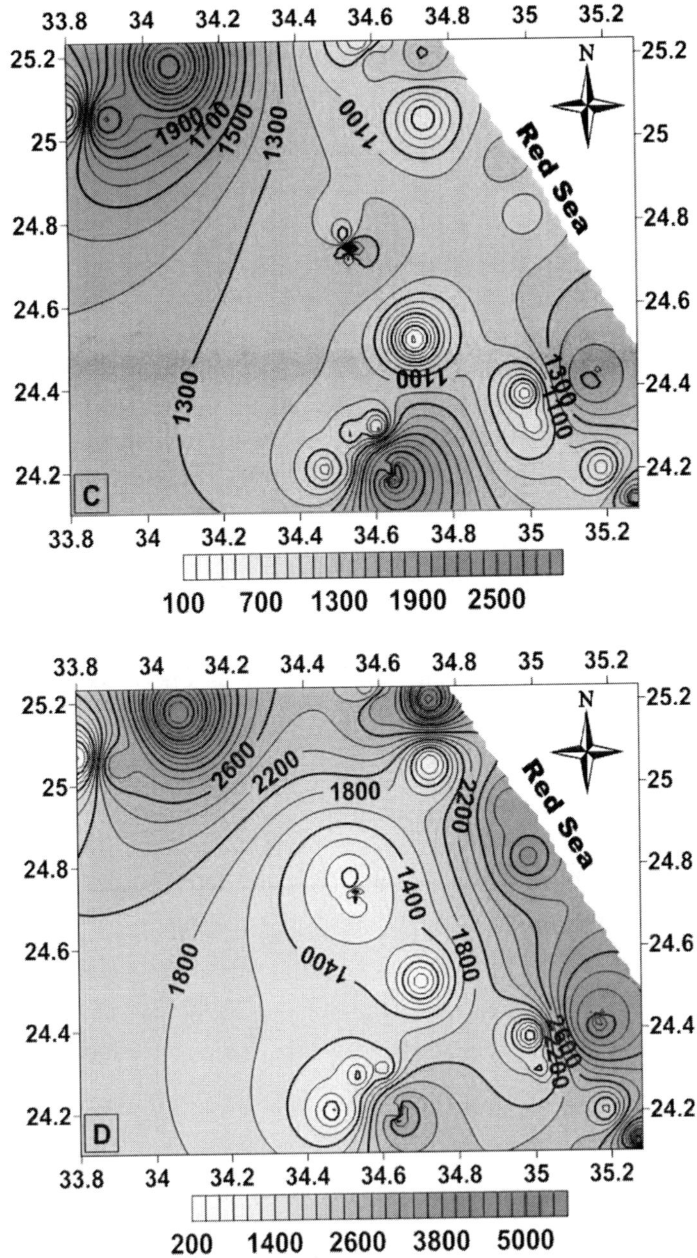

Figure 6. Spatial distribution of TDS and main chemical elements over the study area. (a) TDS; (b) Calcium; (c) Sodium; (d) Chloride.

The spatial distribution of TDS (Figure 6(a)) is quite similar to that of chloride, calcium and sodium. This corroborates the hypothesis that the main contributors to the groundwater salinity are chloride, calcium and sodium.

8. MAIN HYDROCHEMICAL FACIES

In order to properly identify the hydrochemical facies and to yield an indication of the qualitative aspects of groundwater, the graphical representation of the results of analysis proves an unavoidable tool. In this study, classification and interpretation of the groundwater hydrochemistry were carried out using the Piper diagram [24] . To achieve this goal, the hydrochemical data were processed with the use of the software Diagrammes [25] . Figure 7 shows the Piper plot of samples. This diagram clearly shows that the groundwater samples have a dominance of chloride and sulfate ions for anions, while sodium is the cation which marks the majority of samples and is followed by the calcium. This reveals the dominance of chloride-sodium facies, and the secondary chloride-calcium and sulfate-sodium facies.

The main hydrochemical facies which characterizes groundwater in each aquifer rock is as following: sulfate-sodium type in meta-volcanic rocks, chloride-sodium and sulfate-sodium types in granitic rocks, chloride-calcium type in basic-ultrabasic rocks. Groundwater in sedimentary rocks displays a chloride-sodium facies (Quaternary rocks) and a chloride-calcium facies (Miocene rocks).

Samples taken in wells close to the Red Sea are of chloride-sodium type, whatever the aquifer, reflecting the effect of marine intrusion.

9. BINARY DIAGRAMS AND MINERALIZATION PROCESS

Dissolved species and their relationship with each other can reveal the origin of solutes and the processes that generated observed composition of the groundwater [26] - [35] . The relationships between concentrations of major dissolved elements are shown in Figure 8. The Na^+ vs. Cl^-relationship has been often used to identify the mechanism for acquiring salinity. The Na^+ vs. Cl^-plot shows that the samples are not aligned along the 1:1 line, but are quite scattered with a weak correlation (R = 0.54) indicating that Na^+ and Cl^- are not for the most part derived from the same source. Figure 8(a) shows however noticeable enrichments or losses in Na^+ contents in the groundwater considering Cl^- as a conservative element. Molar Na/Cl ratios range from as high as 3.91 to 0.41. High values are found for the meta-volcanic rocks (average = 1.63), the granites (average = 1.14) and the meta-sediments (average = 2.20). The lowest values are found in the sedimentary rocks (average = 0.61). The high Na/Cl ratios in crystalline rocks are probably controlled by water?rock interaction, for example, albite (reaction 3) or analcite (reaction 4) weathering according to the following reactions:

$$2NaAlSi_3O_8 + 9H_2CO_3 \rightarrow Al_2Si_2O_5(OH)_4 + 2Na^+ + 2HCO_3^- + 4H_4SiO_4 \quad (3)$$

$$NaAlSi_2O_6 + 3/2\,H_2O + CO_2 \rightarrow Na^+ + HCO_3^- + 1/2\,Al_2(Si_2O_5)(OH)_4 + H_4SiO_4 \quad (4)$$

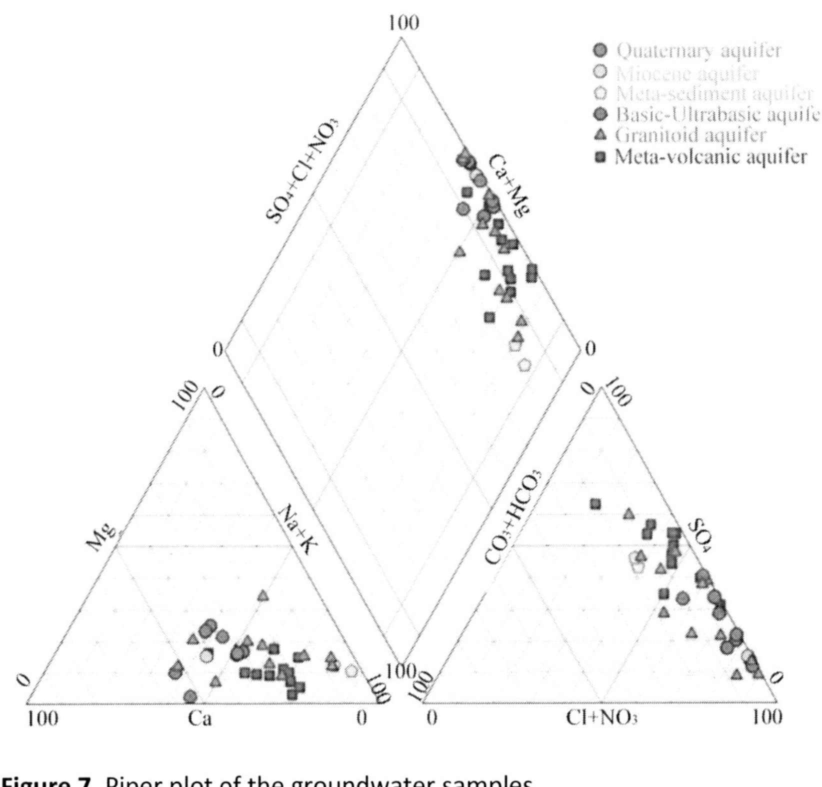

Figure 7. Piper plot of the groundwater samples.

Samples with low Na/Cl ratios are probably the result of Na^+ loss or exchange with soils and clays. As discussed below, the Ca^{2+} and Mg^{2+} data support this idea.

In order to understand the origin of the water salinity due partly to chlorides, the bromide and the chloride ions, which are good conservative tracers in aquifers, are considered. This, because they do not participate in oxydo-reductions reactions and do not form insoluble precipitates [36] . The Br/Cl ratio is used in this case to determine the origin of salinity. Indeed, it is often the most relevant ratio to explain the salinity of groundwater [37] . The study of the Br/Cl ratio allows discriminating the areas under evaporitic influences Seawater has molar Cl/Br ratios of 550 - 700 [38] - [40] . Coastal precipitation has similar Cl/Br ratios, however, inland precipitation, especially in arid or semiarid climates, may have lower Cl/Br ratios due to the tendency for Cl to be removed in early rainfall near coastal areas [38] [39] [41] . The average Cl/Br are high in the crystalline rocks (Meta-volcanic rocks: $\overline{Cl/Br}$ = 1046; Basic-Ultrabasic rocks: $\overline{Cl/Br}$ = 1404; Granite: $\overline{Cl/Br}$ = 1233; Metasediments: $\overline{Cl/Br}$ = 1365). For the sedimentary rocks, the average $\overline{Cl/Br}$ is 733. This value is close to the seawater Cl/Br ratio. This shows the effect of seawater intrusion in the sedimentary rocks (Quaternary and Miocene), which are located close to the seaside. In all crystalline and sedimentary rocks within the study area, the Cl/Br is much higher compared to the Cl/Br ratio of halite, which is about 105 [42] . Thus, halite dissolution must be relatively minor as a cause of increasing salinity in the area.

The excess of Cl^- can be explained by the combined effect of another source for this ion than the dissolution of halite, together with the Na^+ losses due to the phenomenon of base exchanges, as clays in bedrock can release Ca^{2+} after setting Na^+. The excess of Cl^- may also have an anthropogenic origin (wastewater, agricultural fertilizers).

The alkaline earth elements may be derived by dissolution of silicate minerals (e.g., plagioclase feldspar, chlorite, or biotite), carbonates (dolomite or calcite), gypsum, and/or cation exchange of Na^+ for Ca^{2+} and Mg^{2+} on clay minerals. HCO^{3-} is poorly correlated with Ca^{2+} (R^2 = 0.19; Figure 8(b)) and Mg^{2+} (R^2 = 0.002). The excess of Ca^{2+} and Mg^{2+} indicates that dissolution of carbonate rocks (calcite, dolomite) is not the main source for these elements, which is also backed by the very poor relation of Ca^{2+} with Mg^{2+} (R^2 = 0.24). Carbonate deposits are absent in the study area, but carbonate cements may be present.

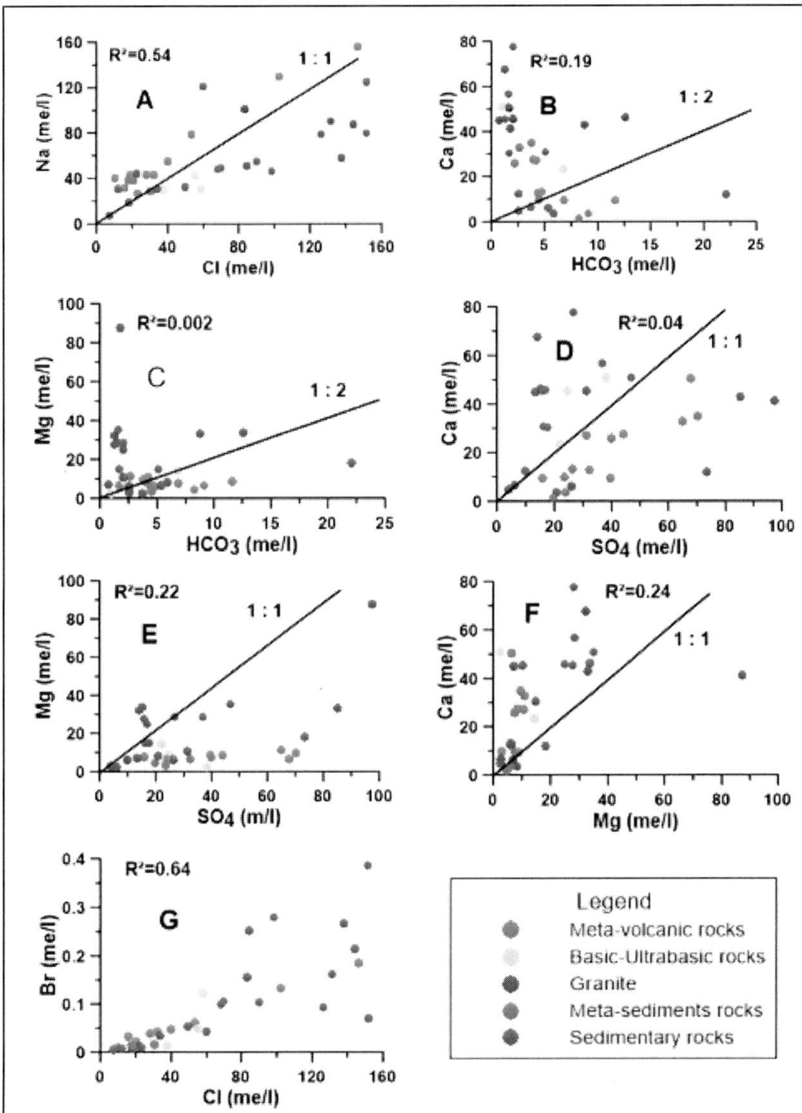

Figure 8. Bi-elements diagrams between the main chemical elements.

Correlation of SO_4^{2-} with Ca^{2+} and Mg^{2+} is also poor (respectively $R^2 = 0.04$ and $R^2 = 0.21$;Figure 8(d) and Figure 8(e)) and shows that these elements (Ca^{2+}, Mg^{2+}, SO_4^{2-}) are not derived by the dissolution of gypsum or any Mg-sulfate mineral.

In Figure 8(d) and Figure 8(e), SO_4^{2-} is in excess compared to Ca^{2+} and Mg^{2+} in water in metavolcanic rocks. The abundance of sulfates could be attributed to the dissolution of minerals as alunite ($KAl_3(SO_4)_2(OH)_6$ (reaction 5):

$$(K,Na)Al_3(SO_4)_2(OH)_6 \rightarrow 3AlO(OH) + (K,Na)^+ + 3H^+ + 2SO_4^{2-} \quad (5)$$

Excess of Ca^{2+} compared to SO_4^{2-} in basic-ultrabasic rocks may be caused by means of hydrolysis of plagioclase feldspars, as Anorthite (reaction 6):

$$CaAl_2Si_2O_8 + 3H_2O + 2CO_2 \rightarrow Ca^{2+} + 2HCO_3^- + Al_2(Si_2O_5)(OH)_4 \quad (6)$$

The plot of Mg^{2+} versus Ca^{2+} (Figure 8(e)) indicates an excess of Ca^{2+}, which can be explained by the existence of geochemical phenomena such as the Ca^{2+} « Mg^{2+} exchanges during the reactions of precipitation or recrystallization of calcite [43] .

Contents of K^+ in waters are very low compared to the contents of other cations (Na^+, Ca^{2+}, Mg^{2+}). Correlation of K^+ with Ca^{2+} and Mg^{2+} is also rather poor, respectively R = 0.27 and R = 0.56. The relative amount of K^+ contained within the clay minerals with respect to the concentrations in groundwater is probably much larger than Ca^{2+} or Mg^{2+}. Hence, the reactions amongst the clay minerals are likely to cause larger relative changes in K^+ than in Ca^{2+} or Mg^{2+}. This may explain the poorer correlation of K^+ with Ca^{2+} and Mg^{2+}, respectively R = 0.27 and R = 0.26.

10. IONS EXCHANGES REACTIONS

The plots of Ca^{2+} vs. HCO_3^- (Figure 8(d)) and Ca^{2+} vs. Mg^{2+} (Figure 8(e)) display a substantial excess of Ca^{2+}, showing that the origin of Ca^{2+} is not the only dissolution of calcite. This is consistent with the hypothesis of a contribution of Ca^{2+} by ion exchange reaction via a basic reaction such as:

$$Ca^{2+} - clay + 2Na^+(water) \rightarrow Ca^{2+}(water) 2Na^+ - clay \quad (7)$$

Na^+ can exchange Ca^{2+} and Mg^{2+} sorbed on the exchangeable sites of the clay minerals, resulting in the increase of Ca^{2+} and Mg^{2+} and the decrease of Na^+ in groundwater. During this process the host rocks are the primary sources of dissolved solids in the water. Groundwater in which the alkaline earths (Ca^{2+}, Mg^{2+}) have been exchanged for the Na^{+} ions is referred to as base-exchange-hardened water [26] (Gupta et al., 2008).

Knowledge of the changes undergone by the chemical composition of the groundwater during its travel in the aquifer is essential. The ion exchange between the groundwater and its host aquifer during travel can be understood analyzing the plot of Ca^{2+} + Mg^{2+} versus $SO_4^{2-} + HCO_3^-$. In this scatter diagram, the points falling along the equiline (Ca^{2+} + Mg^{2+} = $SO_4^{2-} + HCO_3^-$) suggest that these ions have resulted from the dissolutions of calcite, dolomite and gypsum [44] - [47] . If reverse ion exchange is the dominant process, it will shift the points to the left due to a large excess of Ca^{2+} + Mg^{2+} over $SO_4^{2-} + HCO_3^-$. The analysis of the Ca^{2+} + Mg^{2+} versus $SO_4^{2-} + HCO_3^-$. Plot (Figure 9) reveals that all the groundwater samples have undergone the process of ion exchanges. Samples taken in the sedimentary, basic-ultra-basic and some samples in granites fall in the Ca^{2+} + Mg^{2+} side (Figure 9), suggesting that reverse ion exchange is the major hydrogeochemical process operating in these rocks. All samples taken in the meta-sediments and meta-volcanic rocks and a few samples in granites fall in the $SO_4^{2-} + HCO_3^-$ side, suggesting exchanges between earth-alkaline in waters and alkaline in clays. Most samples taken in granitic rocks present excess of Ca^{2+}.

Ion exchange process is further discussed using the chloro-alkaline index (CAI) [48] , where all values are expressed in meq·l^{-1}:

$$CAI = \left(Cl^- - \left(Na^+ + K^+\right)\right)/Cl^- \quad (8)$$

This ratio is positive when the sodium and potassium contents are low, i.e. when groundwater has been in contact with minerals able to release the interchangeable alkaline-earth ions easily. These exchanges of ions contained in the matrix with those present in the groundwater are very variable and depend, among other things, on the nature of the substrate. This ratio is negative when the sodium and potassium contents are high, i.e. when groundwater has been strongly in contact with minerals able to release the interchangeable alkaline ions easily. This phenomenon of base exchanges is most generally known with the alumino-silicated clays formed by layers or sheets whose cohesion is ensured by the existence of interlayer ions and water. Indeed, the surface of the layers is negatively charged, thus promoting the possibility of ion exchange with those of the groundwater.

Thus when there is an exchange between adsorbed Na^{+} or K^{+} with Mg^{2+} or Ca^{2+} in the groundwater, the CAI will be negative and if there is a reverse ion exchange prevalent (exchange between adsorbed Mg^{2+} or Ca^{2+} with Na^+ or K^+ in the groundwater) then this index will be positive. The average CAI (Table 7) is positive for the basic-ultrabasic rocks and sedimentary rocks reflecting the substitution of sodium and potassium in groundwater with calcium and magnesium in the underground environment. In meta-volcanic, granites and meta-sediments rocks, this index is in average negative pointing out gain in Na^+ and K^+, and losses in Ca^{2+} and Mg^{2+}.

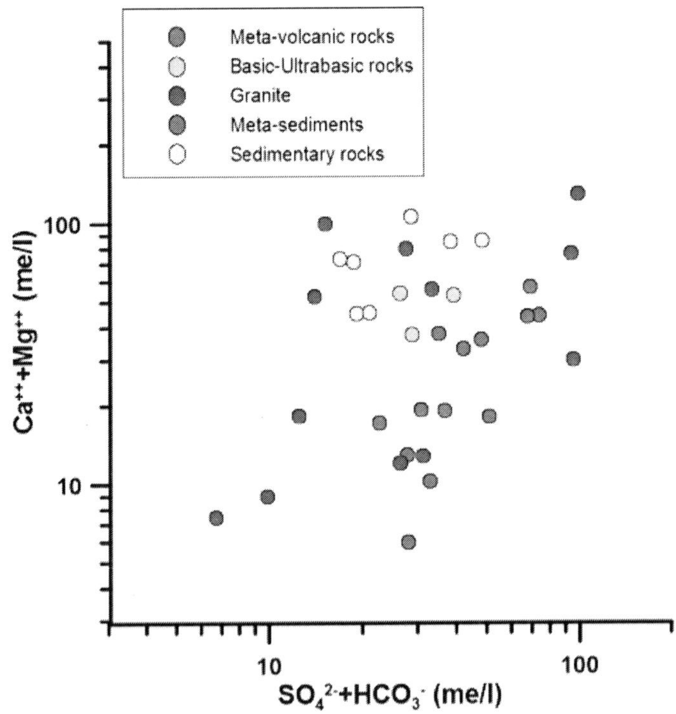

Figure 9. Ca^{2+} + Mg^{2+} vs. $SO_4^{2-} + HCO_3^-$ diagram.

The dominance of positive or negative values of the chloro-alkaline index for each type of rocks corroborates well the relationship previously identified between various elements in solution in the groundwater. Despite the specific hydrogeological conditions of the study area, the chloro-alkaline index (CAI), together with the diagram of Ca^{2+} + Mg^{2+} versus $SO_4^{2-} + HCO_3^-$ allowed to understand that the contact of groundwater with the different aquifer rocks supported the spatial variability of ions by the process of base exchanges.

11. MINERAL SATURATION INDICES

Saturation indices of minerals are very useful for evaluating the extent to which water chemistry is controlled by equilibrium with solid phases (Appelo and Postma, 1993). The saturation indice (SI) was evaluated using the software PHREEQC V2 [49] . PHREEQC can be used via the freeware DIAGRAMMES [25] . The thermodynamic database used to this purpose is Wateq.dat [50] . The saturation indices of minerals that were suspected to be responsible for the chemical composition of the Eastern desert groundwater were computed. PHREEQC uses the specific ionic concentrations in the water and the mass balance approach to calculate all the stoichiometrically available reactions that are responsible for the observed chemical changes between end member waters [51] . The package calculates the saturation indices, SI, of minerals using the concentrations of the major ions in the system. The saturation index of a mineral is obtained from Equation (9) [52] [53] :

$$SI = \log(IAP/KT) \tag{9}$$

where IAP is the ion activity product of the chemical element in solution, KT is the equilibrium constant of the reaction considered at the temperature T(K). When the SI is below 0, the water is undersaturated with respect to the mineral in question. An SI of 0 means water is in equilibrium with the mineral, whereas an SI greater than 0 means a supersaturated solution with respect to the mineral in question.

The saturation indices of the following minerals in all water samples were evauated: calcite, aragonite, dolomite, anhydrite, gypsum and halite. The average values of the SI for each rock type are given in Table 7.

Calculations showed that carbonate minerals have different degrees of saturation. The dolomite SI ranges from −1.18 to +3.90, that of calcite ranges from 0.02 to +1.79. Assuming that equilibrium is in the range of −0.5 to +0.5, the results show that both calcite and dolomite have reached equilibrium in the basic-ultrabasic rocks and are in a state of supersaturation in other rocks.

Table 7. Average Choro-Alcaline Index (CAI) and Saturation Indices (SI) of water samples in aquifer rocks types.

Rocks types	CAI	SI Dolomite	SI Calcite	SI Aragonite	SI Gypsum	SI Anhydrite	SI Halite
Meta-volcanic	−0.64	1.55	0.92	0.78	−0.28	−0.50	−4.57
Basic-ultrabasic	0.29	0.45	0.54	0.40	−0.15	−0.37	−4.58
Granitoids	−0.16	1.66	0.85	0.71	−0.57	−0.78	−4.50
Meta-sediments	−1.24	1.89	0.69	0.54	−1.27	−1.48	−4.91
Sedimentary	0.38	0.88	0.51	0.31	−0.26	−0.48	−4.04

The evaporitic minerals show degrees of saturation lower than the carbonate minerals. Gypsum SI ranges from −1.49 to 0.13 and anhydrite SI from −1.71 to −0.09. Water is in equilibrium regarding these minerals in all rocks, except in meta-sediments where it is undersaturated. Halite SI ranges from −5.94 to −3.52 indicating that the groundwater is very undersaturated regarding this mineral. This results also indicates that halite is not the main source for Na^+ and Cl^-.

12. ROLE OF EVAPORATION

The study region experiences dry and semiarid climatic condition, evaporation may also contribute in groundwater chemistry. The sodium versus chloride ratio can be used to identify the evaporation process in groundwater. Evaporation will increase the concentration of total dissolved solids in groundwater and the Na/Cl ratio remains the same. It is a rather efficient indicative factor of evaporation. If evaporation is the dominant process, Na/Cl ratio should remain constant when TDS increases [54] . Accordingly, the plot of Na/Cl versus TDS would give a horizontal line. The TDS versus Na/Cl scatter diagram of the groundwater samples (Figure 10) shows that the points fall along a horizontal line which means that the (Na^+/Cl^-) remains almost constant despite the increase in TDS. This indicates that evaporation plays a decisive role in controlling the chemistry of groundwater.

13. GROUNDWATER POTABILITY ASSESSMENT

The Eastern Desert groundwater potability was assessed according to international standards (WHO 2008) and to the water hardness. The hardness is expressed in French degrees (°F). A French degree (°F) corresponds to the hardness of a solution containing 10 mg/l of $CaCO_3$. 1°F is equivalent to 4 mg of calcium per liter and 2.4 mg of magnesium per liter. 1meq of calcium ion is equivalent to 5°F [55] .

The contents of the main elements of samples were compared with WHO standards [56] and are reported in Table 8. This table shows that no water sample fully meets the WHO criteria for potability. In each rock type, the contents of several elements exceed WHO standards: metavolcanic rocks (Ca^{2+}, Na^+, Cl^-, SO_4^{2-}, TDS); basic-ultrasic rocks (Ca^{2+}, Na^+, K^+, Cl^-, SO_4^{2-}, TDS);

granites (Na); meta-sediments (Mg^{2+}, Na^+, K^+, Cl^-, SO_4^{2-}, HCO_3^-, TDS), sedimentary rocks (Ca^{2+}, Mg^{2+}, Na^+, Cl^-, SO_4^{2-}, TDS). The TDS limit value is exceeded in all samples except in 8% of samples in granitic rocks. It appears from this table that only the waters in granitic rocks best meet the WHO criteria for potability. The waters of other rocks are rather far from WHO standards.

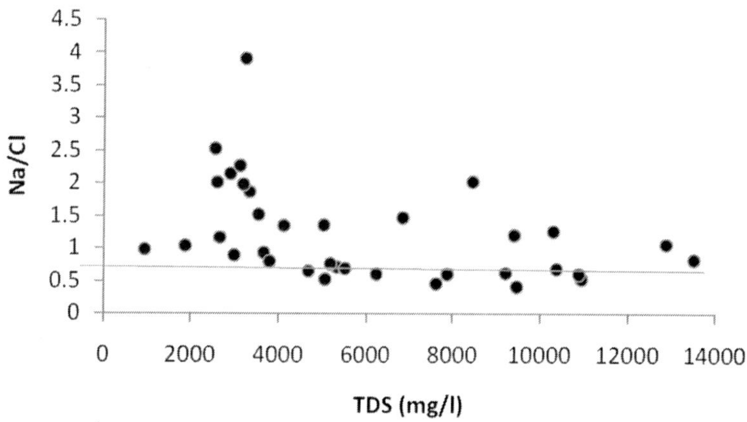

Figure 10. Na^+/Cl^- vs. TDS diagram.

The hardness ranges from 30˚F to 643˚F, showing that some samples are soft while other are very hard. The average hardness is as follows for each rock type: Meta-volcanic = 154˚F, Basic-ultrabasic = 241˚F, granites = 242˚F, Meta-sediments = 41˚F, sedimentary rocks = 365˚F. These figures show that hardness is quite high in all rock types, except in the meta-sediments. The contents of magnesium and calcium that exceed the threshold set by WHO, cause such high hardness in these rock aquifers.

These results show that the groundwater in the study area is of poor to bad quality for drinking purpose.

Table 8. Percentage of samples having ion concentrations below the WHO standards.

Elements	WHO standards	Meta-volcanic (%)	Basic-ultrabasic (%)	Granitoids (%)	Meta-sediment (%)	Sedimentary (%)
Ca^{2+} (mg/l)	100	0	0	12	100	0
Mg^{2+} (mg/l)	50	9	33	16	0	0
Na^{+} (mg/l)	150	0	0	0	0	0
K^{+} (mg/l)	12	63	0	33	0	14
Cl^{-} (mg/l)	250	0	0	58	0	0
SO_4^{2-} (mg/l)	250	0	0	16	0	0
HCO_3^{-} (mg/l)	250	45	66	58	0	85
TDS (mg/l)	1500	0	0	8	0	0
pH	$6.5 < pH < 9.5$	100	100	100	100	100

14. SUMMARY AND CONCLUSIONS

Understanding groundwater hydrochemistry and quality is vital to preserve this resource so that it can meet the present and future water needs in many countries. In the Eastern Desert in Egypt, groundwater resources play a vital role in supplying water for drinking, industrial and agricultural purposes. However, there is a lack of knowledge of the groundwater mineralization processes and a lack of rational management. This study adds to our understanding of the chemical properties of groundwater in this area in Egypt.

This study allowed first for a hydrogeological synthesis of the aquifer system in this part of the Eastern Desert. This system is composed of several rock types including Pre-Cambrian crystalline fractured rocks and Tertiary/ Quaternary sedimentary rocks along the Red Sea border. At the scale of the Eastern Desert, all these formations have a hydraulic continuity and form a single aquifer system. The piezometric map displays an N-S divide line. General groundwater flow takes place on either side of this line, to the East toward Red Sea and to the West towards the Nile River.

The hydrogeochemical study of the groundwater was conducted using several methods (Piper diagram, principal components analysis, ions exchanges, saturation indices of various minerals). This study revealed the presence of different hydrochemical facies in each aquifer rock type: chloride-sodium and sulfate-sodium in granitic rocks, chloride-calcium in basic-ultrabasic rocks, chloride-sodium in quaternary rocks and a chloride-calcium in Miocene rocks. Close to the Red Sea, groundwater is of chloride-sodium type, reflecting the effect of marine intrusion. The groundwater chemical characteristics are controlled by natural geochemical processes but also by some anthropogenic activities. Evolution of the groundwater chemistry is controlled by multiple processes including natural minerals dissolution as halite or calcite. However silicate weathering and ion exchanges play a prominent role in the ion enrichment of groundwater. Seawater intrusion on the Red Sea coastline and anthropogenic activities affect also the groundwater chemistry. The role of evaporation due to high temperatures in this area was also highlighted.

In the Eastern Desert, the majority of chemical elements analyzed exceed the standards set by WHO. The groundwater is accordingly not perfectly appropriate for human consumption.

The results of this study helped to significantly improve the understanding of the aquifer system of Southern Eastern Desert which is an important resource for the development of this region in Egypt and may contribute as well for a better management of this vital resource.

REFERENCES

1. Andre, L., Franceschi, M., Pouchan, P. and Atteia, O. (2005) Using Geochemical Data and Modelling to Enhance the Understanding of Groundwater Flow in a Regional Deep Aquifer, Aquitaine Basin, South-West of France. Journal of Hydrology, 305, 40-62.

2. Rashed, M., Idris, Y. and Shaban, M. (2006) Integrative Approach of GIS and Remote Sensing to Represent the Hydrogeological and Hydrochemical Conditions of Wadi Qena-Egypt. Proceedings of the 2nd International Conference on Water Resources & Arid Environment, Saudi Arabia, November 2006, 1-10.

3. Abd El-Moneim, A.A. (2005) Overview of the Geomorphological and Hydrogeological Characteristics of the Eastern Desert of Egypt. Hydrogeology Journal, 13, 416-425.

4. Said, R. (1990) The Geology of Egypt. A.A. Balkema, Rotterdam/ Brookfield, 734 p.

5. El Ramly, I.M. (1972) Final Report on Geomorphology, Hydrology Planning for Groundwater Resources and Reclamation in Lake Nasser Region and Its Environs. Lake Nasser Center and Desert Institute, Cairo, 603 p. (Unpublished Report)

6. Abdel Kader, A.A. (2001) Application of Some Geophysical and Hydrogeological Techniques for Groundwater Resources Investigation in Selected Areas between Idfu-Marsa Alam, Eastern Desert, Egypt. M.Sc. Thesis, Assiut University, Assiut, 172 p. (Unpublished)

7. Mohamed, A.A. (2004) Geographic Environment and Development in Marsa Alam Area, Eastern Desert, Egypt. M.Sc. Thesis, Cairo University, Giza. (Unpublished)

8. Saleh, M.F. (1993) Hydrogeological and Hydrochemical Studies on Some Localities in South Eastern Desert, Egypt. Ph.D. Thesis, Suez Canal University, Ismailia. (Unpublished)

9. Tahoon, M.A. (2011) Hydrogeochemical and Environmental Study in the Area between Marsa Alam and Baranes, Red Sea Coast, Egypt. Ph.D. Thesis, South Valley University, Qena, 215 p. (Unpublished

10. Embaby, A.I., Razack, M. and Porel, G. (2015) Geophysical Investigations to Highlight Hard Rocks Aquifers Structure in the South Eastern Desert, Egypt. 6th International Conference on Water Resources and Sustainable Development, ENSH, Blida, Algeria, 2015, Abstract 4 p

11. Domenico, P.A. and Schwartz, F.W. (1990) Physical and Chemical Hydrology. Wiley, New York

12. Swan, A.R.H. and Sandilands, M. (1995) Introduction to Geological Data Analysis. Blackwell, Oxford.

13. Gouaidia, L., et al. (2011) évaluation de la vulnérabilitéd'une nappe en milieu semi-arideetcomparaison des méthodesappliquées: Cas de la nappe de Meskiana (Est Algérien). [Vulnerability Assessment of Groundwater in Semi-Arid and Comparison of Methods: Meskiana Groundwater (Eastern Algeria).] Revue Sécheresse, 22, 35-42. (In French)

14. Dawdy, D.R. and Feth, J.H. (1967) Application of Factor Analysis in Study of Chemistry of Groundwater Quality, Mojaveriver Valley, California. Water Resources Research, 3, 505-510.

15. Ashley, R.P. and Lloyd, J.W. (1978) An Example of the Use of Factor Analysis and Cluster Analysis in Ground Water Chemistry Interpretation. Journal of Hydrology, 39, 355-364.

16. Lawrence, F.W. and Upchurch, S.B. (1983) Identification of Recharge Areas Using Factor Analysis. Ground Water, 20, 680-687.

17. Usunoff, E.J. and Guzman, A.G. (1989) Multivariate Analysis in Hydrochemistry. An Example of the Use of Factor and Correspondence Analysis. Ground Water, 17, 27-34.

18. Razack, M. and Dazy, J. (1990) Hydrochemical Characterization of Groundwater Mixing in Sedimentary and Metamorphic Reservoirs with Combined Use of Piper's Principle and Factor Analysis. Journal of Hydrology, 114, 371-393.

19. Abderamane, H., Razack, M. and Vassolo, S. (2012) Hydrogeochemical and Isotopic Characterization of the Groundwater in the Chari-Baguirmi Depression. Republic of Tchad. Environmental Earth Sciences, 69, 2337-2350.

20. Hussein, M.T. (2004) Hydrochemical Evaluation of Groundwater in the Blue Nile Basin, Eastern Sudan, Using Conventional and Multivariate Techniques. Hydrogeology Journal, 12, 144-158.

21. Yitbarek, A., Razack, M., Ayenew, T., Zemedagegnehu, E. and Azagegn, T. (2012) Hydrogeological and Hydrochemical Framework of Upper Awash River Basin, Ethiopia: With Special Emphasis on Interbasins Groundwater Transfer between Blue Nile and Awash Rivers. Journal of African Earth Sciences, 65, 46-60.

22. Taqveem, A.K. (2015) Groundwater Quality Evaluation Using Multivariate Methods, in Parts of Ganga Sot Sub-Basin, Ganga Basin, India. Journal of Water Resource and Protection, 7, 769-780.

23. Harman, H.H. (1960) Modern Factor Analysis. University of Chicago Press, Chicago.

24. Piper, A.M. (1944) A Graphic Procedure in the Geochemical Interpretation of Water Analyses. Transactions—American Geophysical Union, 25, 914-923.

25. Simler, R. (2009) Diagrammes Software. http://www.lha.univ-avignon.fr/LHA-Logiciels.htm

26. Gupta, S., Mahato, A., Roy, P., Datta, J.K. and Saha, R.N. (2008) Geochemistry of Groundwater, Burdwan District, West Bengal, India. Environmental Geology, 53, 1271-1282.

27. Sujatha, D. and Reddy, R.B. (2003) Quality Characterization of Groundwater in the South-Eastern Part of the Ranja Reddy District, Andhra Pradesh, India. Environmental Geology, 44, 579-586.

28. Aboubaker, M., Jalludin, M. and Razack, M. (2013) Hydrochemistry Study of a Volcano-Sedimentary Aquifer Using Major Ion and Environmental Isotope Data. Dalha Basalts Aquifer, Southwest of Republic of Djibouti. Environmental Earth Sciences, 70, 3335-3349.

29. Moussa, A., Zouari, K. and Oueslati, N. (2008) Geochemical Study of Groundwater Mineralization in the Grombalia Shallow Aquifer, North-Eastern Tunisia: Implication of Irrigation and Industrial Waste Water Accounting. Environmental Geology, 58, 555-566.

30. Kuldip, S., Hundal, H. and Dhanwinder, S. (2011) Geochemistry and Assessment of Hydrogeochemical Processes in Groundwater in the Southern Part of Bathinda District of Punjab, Northwest India. Environmental Earth Sciences, 64, 1823-1833.

31. Yuce, G. (2007) A Geochemical Study of the Groundwater in the Misli Basin and Environmental Implications. Environmental Geology, 51, 857-868.

32. Jalali, M. (2009) Geochemistry Characterization of Groundwater in an Agricultural Area of Razan, Hamadan, Iran. Environmental Geology, 56, 1479-1488.

33. Nandimandalam, J.R. (2011) Evaluation of Hydrogeochemical Processes in the Pleistocene Aquifers of Middle Ganga Plain, Uttar Pradesh, India. Environmental Earth Sciences, 65, 1291-1308.

34. Diaw, M., Faye, S., Stichler, W. and Maloszewski, P. (2012) Isotopic and Geochemical Characteristics of Groundwater in the Senegal River Delta Aquifer: Implication of Recharge and Flow Regime. Environmental Earth Sciences, 66, 1011-1020.

35. Singh, S., Raju, N.J. and Ramakrishna, C. (2015) Evaluation of Groundwater Quality and Its Suitability for Domestic and Irrigation Use in Parts of the Chandauli-Varanasi Region, Uttar Pradesh, India. Journal of Water Resource and Protection, 7, 572-587.

36. Fetter, C.W. (1994) Applied Hydrogeology. Prentice Hall Inc., Upper Saddle River, 691.

37. Usho, R., Kur, H., Jessica, S., Ian, E., Michael, N.F., Stephen, J.S. and Andrea, L.H. (2005) The Use of 36Cl and Chloride/Bromide Ratios in

Discerning Salinity Sources and Fluid Mixing Patterns: A Case Study at Saratoga Springs. Chemical Geology, 222, 94-111.

38. Davis, S.N., Whittemore, D.O. and Fabryke-Martin, J. (1998) Uses of Chloride/Bromide Ratios in Studies of Potable Water. Ground Water, 36, 338-351.

39. Davis, S.N., Cecil, L., Zreda, M. and Moysey, S. (2001) Chlorine-36, Bromide, and the Origin of Spring Water. Chemical Geology, 179, 3-16.

40. Petrides, B. and Cartwright, I. (2006) The Hydrogeology and Hydrogeochemistry of the Barwon Downs Graben Aquifer, Southwestern Victoria, Australia. Hydrogeology Journal, 14, 809-826.

41. Herczeg, A.L., Torgersen, T., Chivas, A.R. and Habermehl, M.A. (1991) Geochemistry of Ground Waters from the Great Artesian Basin, Australia. Journal of Hydrology, 126, 225-245.

42. Cartwright, I., Weaver, T.R., Fulton, S., Nichol, C., Reid, M. and Cheng, X. (2003) Hydrogeochemical and Isotopic Constraints on the Origins of Dryland Salinity, Murray Basin, Victoria, Australia. Applied Geochemistry, 19, 1233-1254.

43. Edmunds, W.M., Kay, R.L.F., Miles, D.L. and Cook, J.M. (1987) The Origin of Saline Groundwaters in the Carnmenellis Granite, Cornwall (UK): Further Evidence from Minor and Trace Elements. In: Fritz, P. and Frape, S.K., Eds., Saline Water and Gases in Crystalline Rocks, Geological Association of Canada Special Paper 33, Geological Association of Canada, St. John's, 127-143

44. Datta, P.S. and Tyagi, S.K. (1996) Major Ion Chemistry of Groundwater in Delhi Area: Chemical Weathering Processes and Groundwater Regime. Journal of the Geological Society of India, 47, 179-188.

45. Rajmohan, N. and Elango, L. (2004) Identification and Evolution of Hydrogeochemical Processes in the Groundwater Environment in an Area of the Palar and Cheyyar River Basins, Southern India. Environmental Geology, 46, 47-61.

46. Cerling, T.E., Pederson, B.L. and Damm, K.L.V. (1989) Sodium-Calcium Ion Exchange in the Weathering of Shales: Implications for Global Weathering Budgets. Geology, 17, 552-554.

47. Fisher, R.S. and Mulican III, W.F. (1997) Hydrochemical Evolution of Sodium-Sulfate and Sodium-Chloride Groundwater beneath the Northern Chihuahuan Desert, Trans-Pecos, Rexas, USA. Hydrogeology Journal, 10, 455-474.

48. Schoeller, H. (1977) Geochemistry of Groundwater. In: Brown, R.H., Konoplyantsev, A.A., Ineson, J. and Kovalensky, V.S., Eds., Groundwater Studies—An International Guide for Research and Practice, UNESCO, Paris, 1-18.

49. Parkhurst, D.L. and Appelo, C.A.J. (1999) User's Guide to PHREEQC (Version 2)—A Computer Program for Speciation, Batch-Reaction, One-Dimensional Transport, and Inverse Geochemical Calculations. United States Geological Survey, Water Resources Investigations Report 99-4259, Washington DC, 326.

50. Hounslow, A. (1995) Water Quality Data: Analysis and Interpretation. CRC Press, Boca Raton.

51. Plummer, L. and Back, W. (1980) The Mass Balance Approach: Application to Interpreting the Chemical Evolution of Hydrologic Systems. American Journal of Science, 280, 130-142.

52. Appelo, C.A.J. and Postma, D. (1993) Geochemistry, Groundwater and Pollution. Balkema, Rotterdam.

53. Yidana, S., Ophori, D. and Yakubo, B. (2008) Hydrochemical Evaluation of the Voltaian System. The Afram Plains area, Ghana. Journal of Environmental Management, 88, 697-707.

54. Jankowski, J. and Acworth, R.I. (1997) Impact of Debris-Flow Deposits on Hydrogeochemical Process and the Development of Dry Land Salinity in the Yass River Catchment, New South Wales, Australia. Hydrogeology Journal, 5, 71-88.

55. De Fulvio, S. and Olori, L. (1976) Definitions and Classification of Naturally Soft and Naturally Hard Waters. In: Amavis, R., Hunter, W.J. and Smeets, J.G.P.M., Eds., Hardness of Drinking Water and Public

Health: Proceedings of the European Scientific Colloquium, Luxembourg 1975, Pergamon Press, New York, 95.

56. World Health Organization (2008) Guidelines for Drinking-Water Quality. 2nd Edition, WHO, Geneva. http://www.who.int/ water_ sanitation_health/dwq/2edvol1i.pdf

Chapter 4

Development and Uptake of Scenarios to Support Water Resources Planning, Development and Management – Examples from South Africa

Nikki Funke[1], Marius Claassen[1] and Shanna Nienaber[1]

[1] Natural Resources and Environment Unit, Council for Scientific and Industrial Research, Pretoria, South Africa

1. INTRODUCTION

The international agenda on water resources development reflects societal needs, political agendas, economic realities and the state of resources. The industrial revolution, which started in the 18thcentury, brought social and economic prosperity but also marked a major shift in humanity's impact on the earth's systems. This shift is now referred to as the Anthropocene [1], where humans have brought such vast and unprecedented changes to the planet that this era represents a new geological time interval [2]. Societal needs have shifted since the 1940s from a need for modest food production to a need for increased agricultural productivity that has been met by high yield crops, the use of pesticides, the application of fertiliser and advanced agricultural techniques. This development has averted food shortages, but has also resulted in humanity having to pay a heavy price in terms of increased water use and energy consumption, as well as environmental degradation [3].

From the early 1970s a series of events and key documents has promoted an integrated approach to sustainable development. The 1972 United Nations Conference on the Human Environment considered the need for a common outlook towards the preservation and enhancement of the human environment [4]. The World Commission on Environment and Development advanced this agenda in their report 'Our Common Future', with an emphasis on sustainable development promoting harmony among human beings and between humanity and nature [5]. The International Conference on Water and the Environment that took place in Dublin in 1992 resulted in the development of four guiding principles [6]. These principles, commonly referred to as the Dublin principles, state that: water is a finite resource with economic value and social implications; local communities must participate in water management; water resources management must be developed within a set of policies; and the role of rural populations and women should be recognised. This led to the Rio Declaration and the adoption of Agenda 21, which is a comprehensive plan of action to be implemented globally, nationally and locally in every area in which humanity impacts on the environment [7]. This declaration subsequently became the blueprint for sustainable development world-wide [8].

Uncertainties about societal, economic, political and environmental aspects have proved to be a considerable obstacle to the implementation of sustainable development. Here follow a few examples of such uncertainties. In 1980, the World Development Report of the United Nations [9] estimated that the world population would reach 6.029bn by the year 2000. Five years later, the estimate was updated to 6.088bn [10], with further updates at five yearly increments resulting in estimates of 6.194bn and 6.123bn [11, 12]. The actual population in the year 2000 turned out to be 6.188bn [13]. Future economic development is also uncertain, with the annual growth in Figure 1 showing how the world average varies significantly between years and also how the growth of individual countries (South Africa in this case), does not necessarily follow the global trend and is even more variable between years.

Environmental conditions also vary significantly over time and space, with Figure 2 illustrating the annual deviation of rainfall over southern Africa. This uncertainty is exacerbated by climate projections, which suggest that freshwater resources are vulnerable and have the potential to be strongly impacted by climate change, with wide-ranging consequences for human societies and ecosystems [14].

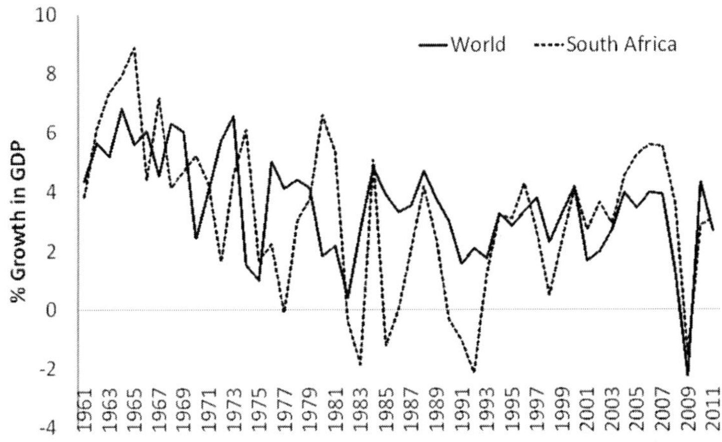

Figure 1. Annual economic growth between 1961 and 2011 [13]

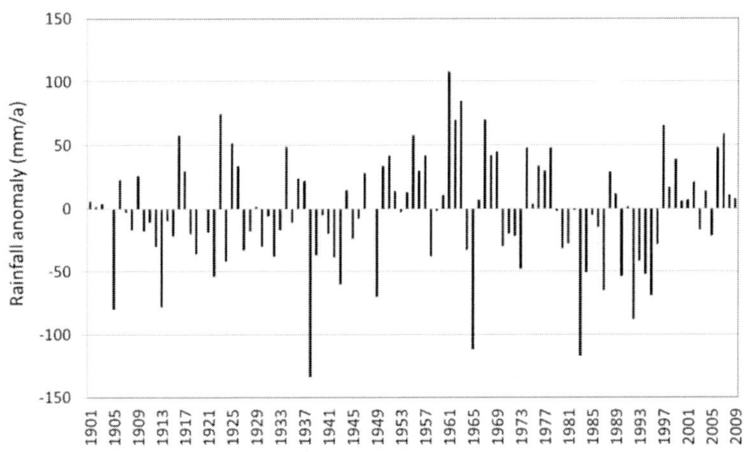

Figure 2. Annual rainfall anomalies for the southern African region (1901-2009; Adapted from [**15**]).

A question that emanates from the realisation that we live in a changing world where change is unpredictable is, 'How do we plan for the future?'

Water use in South Africa was first regulated through the Irrigation and Conservation of Waters Act (Act No. 8 of 1912), which managed the use of water from public streams for domestic, irrigation and industrial purposes [16]. The Water Act (Act No. 54 of 1956) further regulated water use by providing for the control of water pollution and the more effective protection of water resources. The variable distribution of water required the development of infrastructure to capture, store and distribute water. The subsequent expansion of mines, industries and urban areas created a demand for further infrastructure development. When this demand further increased and the social and economic issues in South Africa became increasingly complex in the 1990s as the country was transitioning from*apartheid* to democracy, a shift in thinking was required. As it became clear that engineering solutions to increase water supply were not sustainable, a holistic strategy to meet future needs became more popular [17].

The new National Water Act (Act No. 36 of 1998) [18] emphasised water resources management at national and catchment scales, made specific provisions for the protection of water resources, established mechanisms to ensure equitable and efficient water use and promoted participatory management. The National Water Resources Strategy [19] addressed the balance of future water supply and demand by establishing scenarios. The demand scenarios were based on population growth by 2025, with the high population scenario at 54 million people and the low population scenario at 50 million people. It also established economic growth scenarios, with the upper scenario assuming 4% growth in GDP and the less favourable scenario assuming 1.5% [19].

While there has been much progress in water infrastructure development for services (public benefit), the backlog in issuing water use licenses (mostly for private benefit) stood at 4 318 in 2011 [20]. The protection of water resources has suffered as a result of the government's drive to achieve social and economic development, with South Africa ranked 128 out of 132 countries in the Environmental Performance Index [21]. The National Water Act provides for a balance of responsibilities, ranging from the Minister and Director General at the national level, to Catchment Management Agencies (CMAs) at the basin level and Water User Associations (WUAs) at a sub-basin level. Progress has been slow as after 14 years after the promulgation of the Water Act, only two CMAs (out of the 19 intended) have been established [20]. It can be argued that many hurdles have to be overcome to fully realise cooperative governance for Integrated Water Resources Management (IWRM), with inadequate human and institutional capacity being one of the main factors limiting the efficient management of water resources in South Africa [22]. To illustrate this point: the country's Department of Water Affairs (DWA) reported having 4 286 people in its employment in September 2010, while 1 155 posts were vacant at the time [20].

From the discussion above it becomes clear that we live in a world with social, economic and environmental conditions that are variable and difficult to predict, and the water sector is no exception. This uncertainty provides a challenging environment for policy and institutional development. Scenarios are one way of attempting to achieve a desired outcome in an uncertain and variable future [23]. The rest of this chapter will examine the research question, 'How are scenarios able to achieve impact in an uncertain world, with a particular focus on water resources planning, development and management?' The body of this chapter focuses on the research method, presents an overview of scenario development and the importance of scenario development and how they facilitate more effective water resources planning, development and management, focuses on a few select South African scenarios and the impact they have had and then turns to discussing the impact of scenarios in general. The conclusion wraps up the learning from this chapter and suggests a way forward in terms of future research and designing scenarios for impact.

2. THE ABILITY OF SCENARIOS TO ACHIEVE IMPACT IN AN UNCERTAIN WORLD WITH A FOCUS ON WATER PLANNING, DEVELOPMENT AND MANAGEMENT

2.1. Method

The authors of this chapter conducted an exploratory study on the ability of scenarios to achieve impact in an uncertain world, with particular reference to water planning, development and management. They conducted a review of scenario planning literature in the water and other sectors, and also considered literature focusing specifically on the impact of scenarios. The authors also considered literature on the impact of scientific research and on the science-policy interface. This was accompanied by a search of major databases (e.g. Google Scholar, EBSCO Host and Scopus) to determine where and how the four scenarios discussed in this chapter have been cited. In addition, the authors interviewed selected stakeholders in the water and other sectors who are likely to have been exposed to scenarios and who may use scenarios when making decisions in their workplace.

2.2. Scenarios and Their Importance In The Water Sector

2.2.1. The History Of Scenario Development

The concept of scenario planning has its origin in military applications, with the US Air Force developing 'scenarios' of what the enemy might do and preparing alternative strategies. It was thus aimed at achieving a desired outcome in an uncertain future [24]. At the end of the 1940s, researchers at the RAND Corporation started to investigate the scientific use of expert opinion in planning for the future [25]. The Royal Dutch Shell company employed scenario tools to good effect in the 1970s, when they improved their size and profitability by being prepared to act quickly during the oil price shock of 1973 [26]. In summarising definitions of scenarios, scenarios can be described as a narrative description of a possible state of affairs or development over time, that they are useful to communicate speculations about the future to promote discussion and feedback, and that they can dramatise trends and alternatives, explore the impacts and implications of decisions, choices and policies, and provide cause-and-effect explanations [24].

Clem Sunter is credited with popularising the use of scenarios in South Africa, with 'The World and South Africa in the 1990s', which describe the 'High Road' and 'Low Road' scenarios [27]. The publication was based on work from Anglo American Corporation teams in London and Johannesburg. Subsequently, Adam Kahane facilitated a process that became known as the Mont Fleur scenario project, which was launched in 1992. It explored the question of 'What will South Africa be like in the year 2002?' These scenarios were arrived at collaboratively by a very broad group [28]. The Department of Arts, Culture, Science and Technology (DACST) also deployed scenarios and technology foresighting in the development of South Africa's National Research and Development Strategy, with Kahn initiating and leading the development of the South African National Research and Technology Foresight Project [29]. The Dinokeng team [30] developed '3 Futures for South Africa', which characterised future scenarios based on the effectiveness of the state and the engagement of society. Some of the recent scenario projects in the water sector include the World Business Council for Sustainable Development report on 'Business in the World of Water: WBCSD Water Scenarios to 2025' [31], and the Global Research Alliance (GRA) report on 'Science and Technology-based Water Scenarios for sub-Saharan Africa' [32].

2.2.2. The Importance Of The Use Of Scenarios In Water Resources Planning, Development And Management

Scenarios are important and useful to water resources planning, development and management in a number of ways. In the South Africa context, in particular, scenario development processes have been instrumental in initiating strategic conversations among scenario workshop In the South African context, (e.g. the transition from *apartheid* to democracy), and have helped develop a common language among people with widely divergent views [28]. Those involved in scenario development processes may be inspired to think more broadly about the future and the forces creating it. They may also realise how their particular actions may help to create a desired future [33]; and they may have suggestions about which options exist to direct target audiences on to a desirable path [28]. The knowledge that scenarios generate can therefore potentially empower role players in the water sector and other sectors to engage in participative governance by equipping them with insights into potential futures they may face, and making them aware of the implications of certain decisions, behaviours and actions [23]. Finally, the advantage of communicating scenarios as stories is that they have the psychological impact that other more academic means of communication, for example, graphs and equations, lack. Stories give order and meaning to events, which is crucial for imagining future possibilities [34].

2.3. Some South African Scenarios: Overview And Impact

The discussion in this chapter and the research question were inspired by the development of the Water Sector Institutional Landscape by 2025 scenarios. These scenarios were the main output of a research project led by the authors. In particular, the authors are interested in how these scenarios could be used by potential end-users. Given this question and the importance and potential usefulness of scenarios in facilitating decision-making in a context of uncertainty, it becomes important to reflect on some examples of scenarios that have been developed in South Africa at different points in history and to learn from the impact they have had on different sectors, including the water sector. These scenarios are discussed in chronological order. The section starts with the High Road/Low Road scenarios that were developed late in the *apartheid* era and on the cusp of South Africa's transition to democracy. Secondly, the Mont Fleur scenarios, which were developed during the democratic negotiations, are discussed. Thirdly, the section focuses on the Dinokeng scenarios that were developed in 2009, the year a new president came to power and a serious economic crisis shook the world. The section concludes with the Water Sector Institutional Landscape scenarios that focus on potential futures of the South African water sector in 2025.

2.3.1. High Road/Low Road

2.3.1.1. Overview and Process

The High Road/Low Road scenarios were an initiative by the Anglo American Corporation in the early 1980s and aimed to look into some less conventional approaches to business planning and future investment decisions, given the international economic turbulence of the 1970s and the resultant slump in commodity markets. During this time, South Africa's economic performance was poor and several events resulted in the country becoming increasingly isolated and the government resorting to a rule of force. Careful and gradual reforms by the *apartheid* government in the middle to late 1980s and increasing attempts by members of the white establishment to reach out to black leaders in exile, led to the eventual unbanning of the African National Congress (ANC) and the release of Nelson Mandela in 1990 [35].

The scenarios involved a large-scale exercise with numerous contributors, notably Pierre Wack and Ted Newland, as well as Clem Sunter. Most of the effort went into developing global scenarios which were based on the analysis of key 'drivers' (for example, demography, technology and societal values) of developments in Japan, the USA and USSR (then regarded as the main players of the world economy), and also the ingredients for success of 'winning' nations and world class companies. This work then provided the basis for the South African scenarios. In essence, these scenarios focus on the choice the country was facing to either (through consultation and negotiation) travel on the 'High Road' to a non-racial democracy and increasing prosperity, or, to continue on the 'Low Road' of confrontation, conflict and falling incomes (as a repressive, centralised society and controlled economy) and ending up as a 'waste land' [35].

The scenarios conclude with the need for a 'common vision' to help launch South Africa into the more desirable 'High Road' scenario. This common vision entails putting South Africans first (looking beyond different races and groups), to turn the country into a 'winning' nation and to work towards achieving a certain income per head, all of which would be reached through negotiation [28].

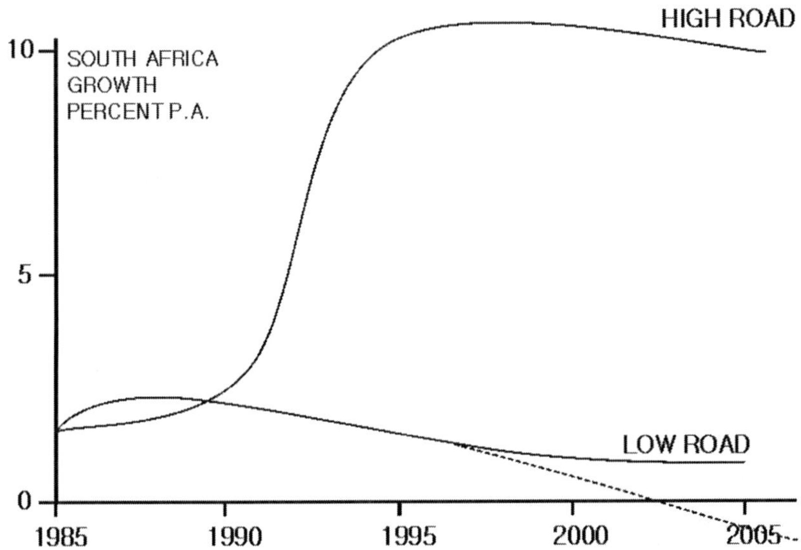

Figure 3. The High Road/Low Road scenarios depicting two possible future trajectories for South Africa [27]

2.3.1.2. Dissemination and Impact

Within a year, starting in 1986, Clem Sunter presented the 'High Road/Low Road' scenarios to 230 (mostly white) audiences at various levels of society, thereby reaching between 25 000 to 30 000 people [35]. Senior politicians of the ANC were also one of Sunter's audiences before the eventual negotiated settlement was reached [28]. The message of the scenarios seems to have made a big impression on the audiences as it was ultimately positive and encouraged people in the country to take it into their own hands to get on to the 'High Road', without being prescriptive about how this should be done [35]. In particular, the High Road/Low Road scenarios also seem to have contributed somewhat to the shift in thinking in government circles, and indeed as supporting evidence for a need for change, which eventually brought about a political transition. In conclusion then, the High Road/Low Road scenarios started out as a corporate scenario project and resulted in a brilliant communication exercise, both in terms of content and style of presentation, that reached thousands beyond the initial intended audience and paved the way for more prominent South Africa scenario exercises to come [35].

In terms of uptake in the scientific and decision-making community, Clem Sunter's book 'South Africa and the World in the 1990s' has been widely cited and includes discussions of a range of topics. These include reflections on various elements of the political and economic transformation of South Africa, the future of Africa, scenario development and planning and globalisation. The citations include a variety of different sources, including books, journal articles, theses and reports. These sources are mostly from the economic, management and social sciences, but also from the health and environmental sciences.

While no examples could be found of the use of the High Road/Low Road scenarios in the water sector, it is likely, judging from the fact that Sunter presented these scenarios to such a wide range of audiences, that some members of government and other stakeholders in the water sector would have been exposed to them in the late 1980s or early 1990s. South Africa's new water legislation certainly reflects the thinking associated with the High Road scenario, with emphasis on introducing ground-breaking new principles into the governance of South Africa's water resources. Though somewhat outdated now, the High Road/Low Road scenarios serve as a reminder of where South Africa could be headed at any point in history. In terms of water resources, South Africa is in need of thoughtful planning, development and management if its water resources are to continue to meet the needs of its ever growing and developing population.

2.3.2. Mont Fleur

2.3.2.1. Overview and Process

The Mont Fleur scenarios were developed in South Africa between 1990 and 1994. Key events during this time were the release of Nelson Mandela, and the legalisation of the ANC, Pan African Congress (PAC) and South African Communist Party (SACP) [36]. The country's first racially inclusive elections were also held at this time. Given this political climate, multiple forums emerged that brought a broad range of stakeholders together to try to develop a new way forward for South Africa. In particular, issues such as housing, education, and constitutional reform received attention [35, 36].

The Mont Fleur scenarios formed a part of this process and essentially tried to encourage debate, thinking and imaginative ideas around how to shape the first ten years of the 'new' South Africa and also to illustrate how certain choices would steer the country towards different outcomes. The Mont Fleur scenario team was made up of a diverse group of 22 prominent South Africans, including politicians, activists, academics and business people [36].

The Ostrich scenario represents a continuation of the *status quo* in South Africa and suggests that no negotiated settlement would be reached and that government would continue to be non-representative [37].

The Lame Duck scenario suggests a South Africa where a settlement would have been achieved but where the transition to a new dispensation would be slow and indecisive [37].

The Icarus scenario suggests a rapid transition to a new government that would push for populist and unsustainable economic policies [37].

The Flight of the Flamingos scenario depicts a government that would choose sustainable policies that would lead the country towards inclusive growth and a maturing democracy [35, 36, 37].

By means of a process of negotiation and reflection on different drivers and concerns, the Mont Fleur scenario team was able to articulate a range of potential outcomes for South Africa during the 1992 to 2002 period. This also helped to clarify the goals and aspirations related to where the country should be heading.

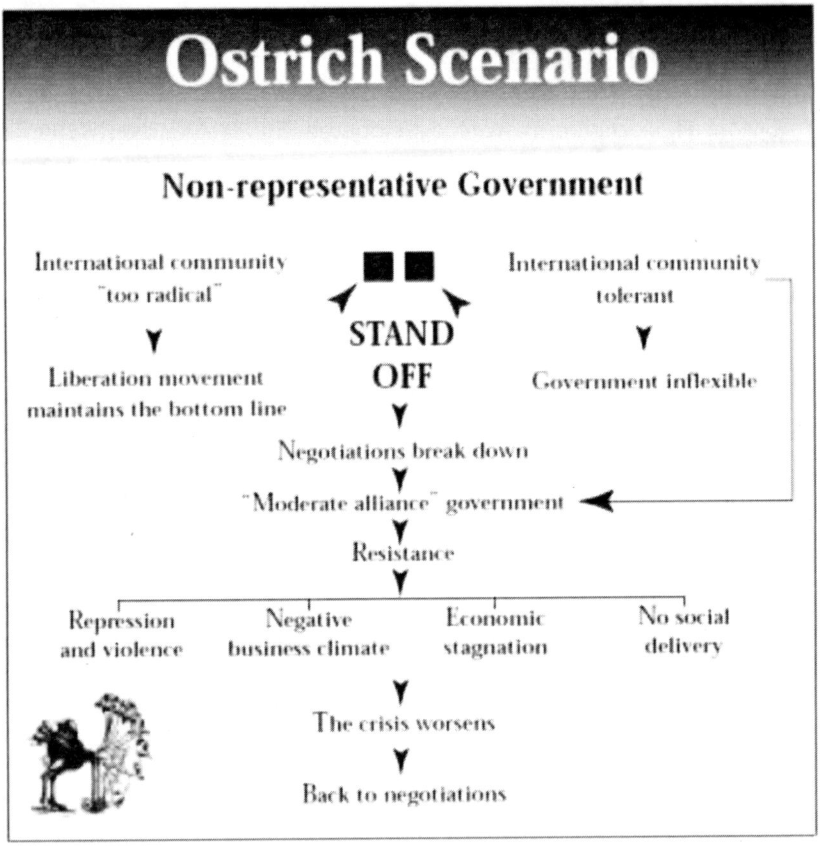

Figure 4. The Ostrich scenario [36]

2.3.2.2. Dissemination and Impact

A variety of dissemination techniques were used by the Mont Fleur team. Key to this process was the fact that each of the individual participants took responsibility for spreading the message of these scenarios. They did this by presenting and discussing the scenarios with more than 50 different groups of people including political parties, companies, academics, trade unions and civil society organisations [36]. This was possible given the diverse background that the team came from. Over and above this process, the scenarios were condensed into an easily accessible 14 page document. This document was distributed to national newspapers. A short video was also produced that combined cartoons with presentations by team members [35, 36].

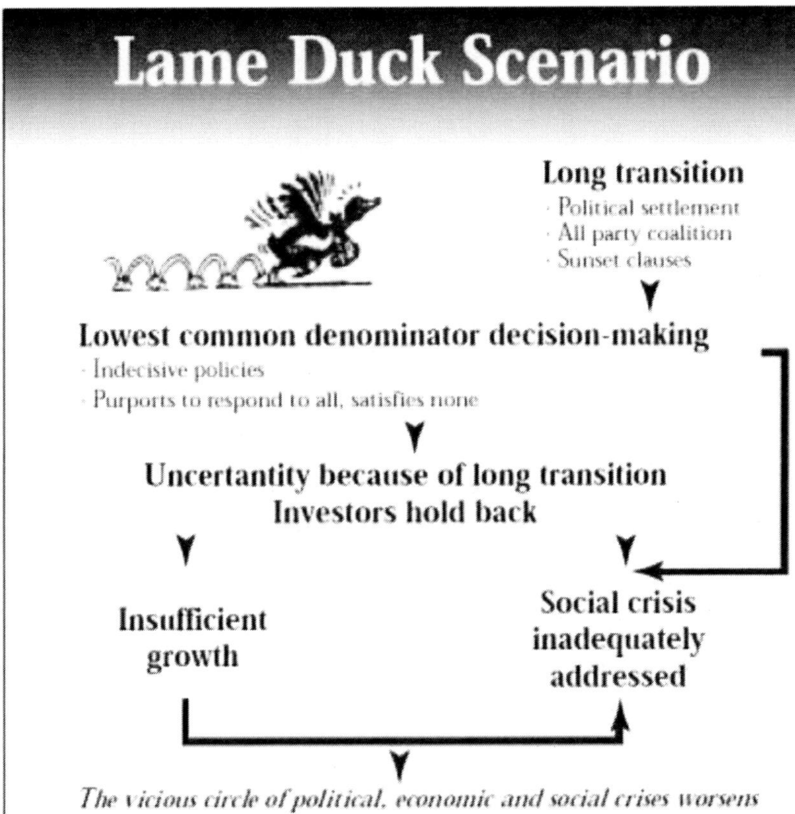

Figure 5. The Lame Duck scenario [36]

The impacts of this scenario development process are subtle. Three key points are important in this regard. Firstly, Mont Fleur, along with other processes taking place in South Africa at the time, helped to establish a common language and understanding about the challenges facing the country and the way forward. This was because participants focused on an issue of common concern for all: 'the future of South Africa'. Secondly, although participants could not agree on one major solution to South Africa's problems, they could agree that certain solutions would not work (such as armed revolutions, continued minority rule and socialism). Thirdly, through an informal process of open conversation, participants who had not expected to agree with each other found common ground and shared understandings about the future of the country [36]. Given these points it is clear that the impact that the Mont Fleur scenarios had was first and foremost on the individuals who participated in the process. There was subsequently a more indirect impact on broader society once these individuals started presenting the scenarios to their various constituencies. Given the widely publicised nature of South Africa's political transition, these scenarios also gained popularity overseas [35].

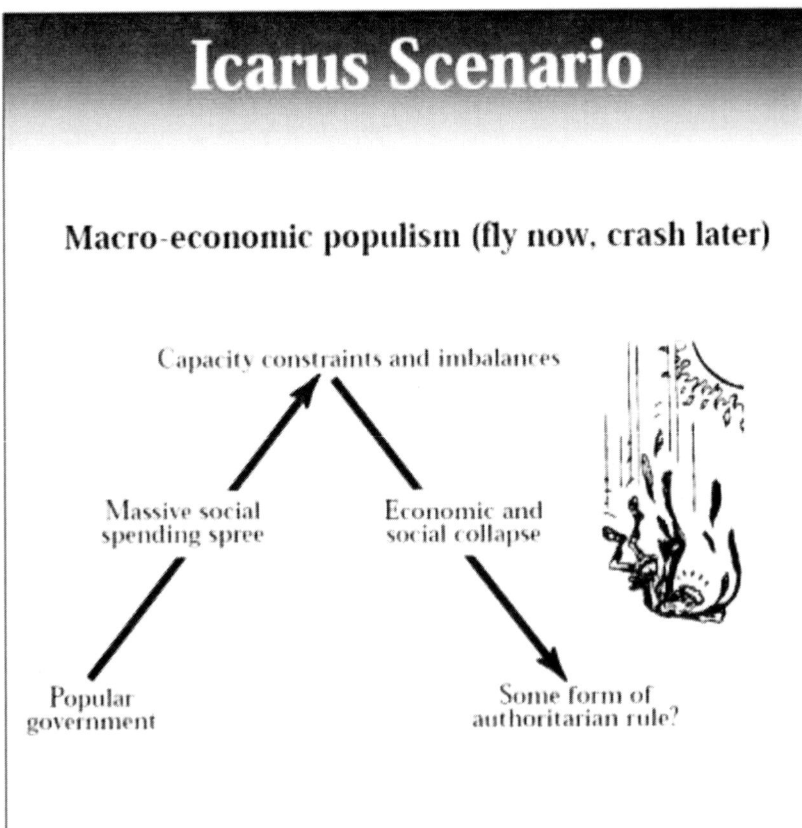

Figure 6. The Icarus scenario [36]

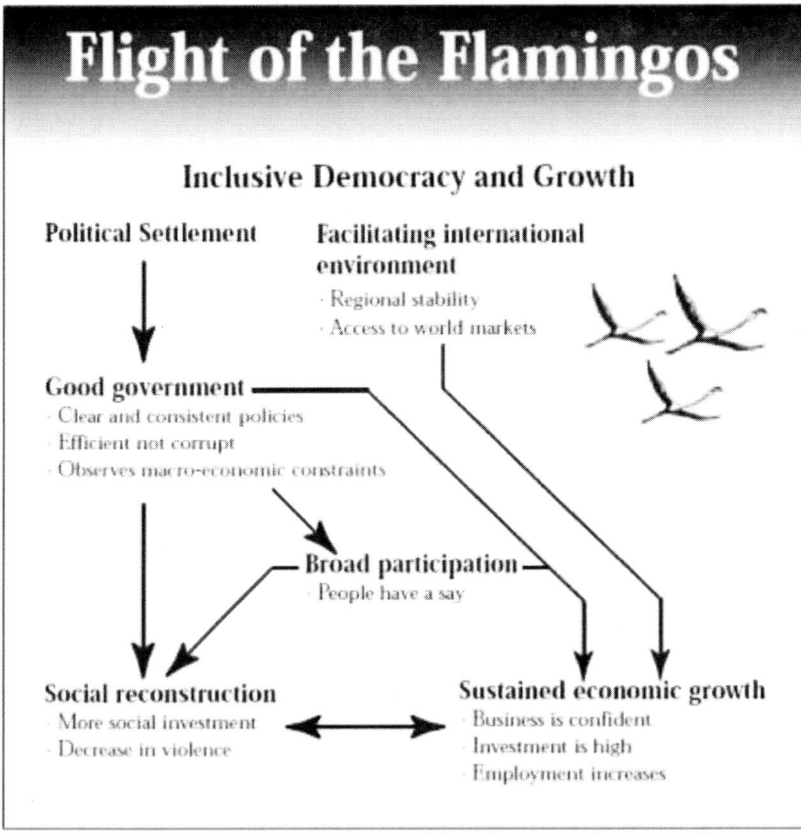

Figure 7. The Flight of the Flamingos scenario [36]

The Mont Fleur scenarios have also been cited in a range of publications. These citations occur in journals, books, conference papers, dissertations and magazines that focus on a range of different disciplines, namely the social, natural and technical sciences. Given this broad interest, the publications cover a broad range of topics most of which are geared towards futures research, democratic transition and strategic planning. This citation record illustrates that the Mont Fleur scenarios seem to have had a considerable impact on the academic community.

Whilst the Mont Fleur scenarios are not obviously related to the South Africa water sector, they did contribute to setting a precedent for using scenario development for planning purposes in South Africa. So, for instance, as mentioned above, the National Water Resources Strategy established a set of water demand scenarios. As with the High Road/Low Road scenarios, the Mont Fleur scenarios were part of the thinking and move towards democratic transition in South Africa. As a result of and in order to complement this change, the water sector was fundamentally transformed and restructured.

2.3.3. Dinokeng Scenarios

2.3.3.1. Overview and Process

The Dinokeng scenario team consisted of 35 leaders from civil society, government, business, political parties, public administration, trade unions, religious groups, academia and the media. The scenario development process was sponsored by the financial institutions Old Mutual and Nedbank who believed that, 15 years into South Africa's democracy, it was important to initiate a reflective and constructive debate about the country's future. According to the Dinokeng scenario team, some of the most prominent challenges facing South Africa are unemployment and poverty, safety and security, education and health. These challenges appear all the more grave in the context of a volatile global economic market, and a global economic crisis that shook the world when these scenarios were developed in 2009 [30].

The Dinokeng scenario team agreed that South Africa needs to realise that the country has failed to appreciate or understand the imperatives of running a modern democratic state, and that there is a problem with the country's self-interested, unethical and unaccountable leadership across all sectors. Additional problems include a weak state that is increasingly less capable of addressing the country's critical challenges, and a population that is either not interested and is showing a growing dependence on the state to provide for everything, or has become co-opted into government or party structures since 1994 [30].

The scenario team developed three possible scenarios which the country could be heading into:

Firstly, the Walk Apart scenario suggests the state becoming increasingly weak and ineffective, and the population, which is looking out for its own interests, eventually losing patience with the state and resorting to protest and unrest to make its views heard. Because the state is unable to meet the population's demands and expectations, it responds brutally, and the result is a spiral of resistance and repression. The Walk Apart scenario therefore suggests a need for South Africans to address their critical challenges, to build state capacity and to organise themselves to engage government in a constructive way, in order to prevent themselves from heading towards disintegration and decline [30].

Secondly, the Walk Behind scenario suggests the state becoming increasingly confident and strong in terms of leading and directing development, fuelled by the fact that civil society is becoming more and more dependent and compliant. The problem is that the state does not have the capacity to address the critical challenges the country is facing on its own. The message of this scenario is that state-led development cannot be successful if there is insufficient state capacity. Furthermore, if the state intervenes constantly and dominates all other sectors, it will crowd out private business and civil society initiatives and will end up creating a population that is complacent and dependent on the state [30].

Thirdly, the Walk Together scenario suggests the state becoming collaborative and increasingly listening to its citizens and leaders from different sectors, engaging with critical voices, and consulting and sharing authority in order to work towards long-term sustainability. In this scenario there is also a focus on a population that takes leadership and holds government accountable and shows an active interest in policy development and outcomes. It is important that South Africans re-engage, that the capacity of the state is strengthened and that leaders from all sectors think beyond their own self-interest and contribute to nation-building [30].

In conclusion, the present contains the seeds for all three futures to be realised. For a healthy democracy and strong socio-economic development to persist, it is important to have a healthy interface between an effective state and an alert and involved population; the nature of this interface is likely to determine the future of the country [30].

2.3.3.2. Dissemination and Impact

In terms of dissemination and impact, once the Dinokeng scenarios on possible futures for South Africa had been developed, the messages of these scenarios were shared with a range of stakeholders. This engagement was followed up with a media and engagement campaign to communicate the Dinokeng scenarios to a variety of organisations, groups and communities across South Africa [30]. The Dinokeng scenarios and the process around their development were also placed on the Dinokeng scenarios website, which is a user-friendly resource for those who are interested in finding out more about these scenarios. The Dinokeng scenarios text is also available for download here.

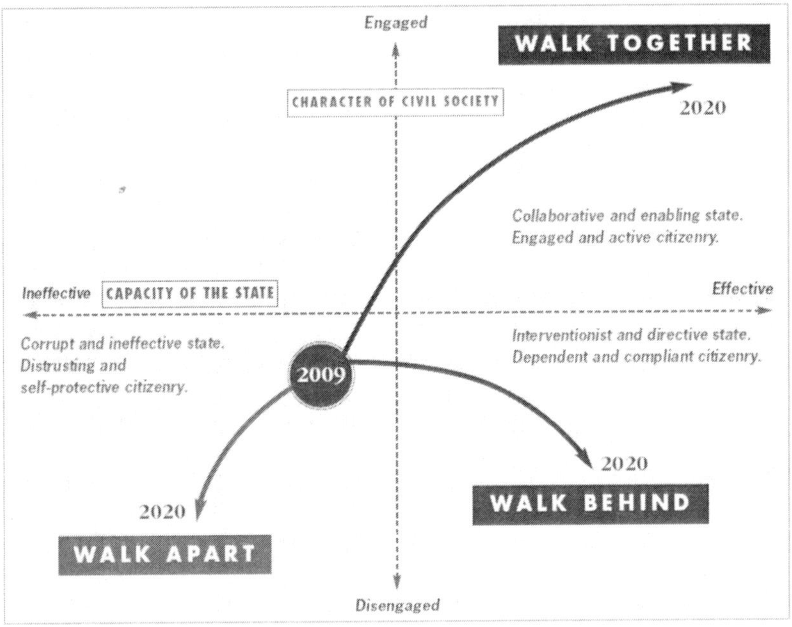

Figure 8. The Dinokeng scenarios [30]

A database search showed that these scenarios have been cited in a wide range of publications. These publications include discussion papers, theses, conference presentations, books and journal papers. The topics of the publications that cited the Dinokeng scenarios are wide-ranging and include issues around interrogating and addressing social issues related to South Africa's democracy, such as local government, education, housing, poverty, unemployment and food security. Many of these topics have a future-centred focus, e.g. investigating South Africans' perceptions about the future, or planning for the future in local government structures. The fact that the Dinokeng scenarios were cited in different kinds of publications and across different subject matters indicates that, at least among the research community, the scenarios were widely distributed and taken up by researchers from different social science-based backgrounds and interests.

A question that arises here is to what extent the Dinokeng scenarios may be of relevance to water resources planning, development and management. While no examples of their use in the water sector were found, it can be argued that the insights provided by these scenarios would prove valuable in focusing on resolving some of the water governance related issues South Africa is currently facing. Examples include problems around water pollution resulting from ineffective waste water treatment and mine and industrial effluents, and water service delivery to previously disadvantaged communities. Those who need to address these water governance related problems could benefit from taking into account the need for maintaining a balance between strong and effective leadership in all sectors and an interested and engaged population, and reflecting on the different future directions such a relationship or lack thereof could take.

2.3.4. Water Sector Institutional Landscape By 2025

2.3.4.1. Overview and Process

An example of scenario development with particular reference to the South African water sector is the South African Water Research Commission's (WRC) Water Sector Institutional Landscape by 2025 scenarios, developed by the authors in 2011 with the assistance of Chan tell Illbury as facilitator, and in consultation with a range of water sector related experts and stakeholders. The focus of the scenario development was on water resources management in South Africa, also with relevance to the water services sector. The aim of these scenarios was to build knowledge about key drivers and uncertainties that relate to the future of the South African water sector, and specifically about the context in which water institutions may operate in future [23].

The knowledge for this project was generated through a structured research process to target existing and new institutional structures and to ensure the involvement and participation of a broad range of stakeholders. The aim of this engagement was to identify water-related needs, priorities and uncertainties based on a wide range of perspectives. A broad range of methods was employed to include stakeholders from both rural and urban environments and with different cultures and educational backgrounds. These included interactive workshops, semi-structured interviews, and a web-based survey. This process was characterised by continuous assessment, learning and adaptation [23].

The key drivers and uncertainties that were identified were subsequently translated into different scenarios that hold potential implications for social and economic development, as well as water resources and services in South Africa. The four scenarios were derived from a matrix with two axes that represent the ability of the decision-making paradigm of water institutions to deal with 'complexity' (refer to the x-axis of the diagram), and the reconciliation of environmental, social and economic demands of present and future generations (referred to as 'sustainability' on the vertical or y-axis of the diagram) [23].

Four possible scenarios emerged from the matrix. The Greedy Jackal scenario depicts a South Africa where water is scarce but government still struggles to meet developmental demands and address backlogs. Under these urgent socio-economic circumstances, environmental responsibility is not prioritised. Despite this the need for a multidisciplinary response to complex water challenges is acknowledged [38].

The Wise Tortoise scenario suggests that a paradigm shift has occurred resulting in a water sector that is multi-layered and engages many different sectors given the strategic importance of the resource in all facets of development. This approach allows for proactive management rather than crisis response to challenges [38].

The Busy Bee scenario suggests that the water sector is defined by great intentions but does not follow up on these with necessary actions. Thus, whilst rhetoric embraces sustainability, in practice there is limited economic and social development to support this process. Part of the challenge is a lack of civil society engagement and failure to embrace the complexity facing water resources management [38].

The Ignorant Ostrich scenario suggests that government fails to recognise water as central to development. As such they rush to implement politically appealing but imbalanced and short term solutions. Civil society is not engaged in decision-making and the complexity inherent to the water sector is overlooked [38].

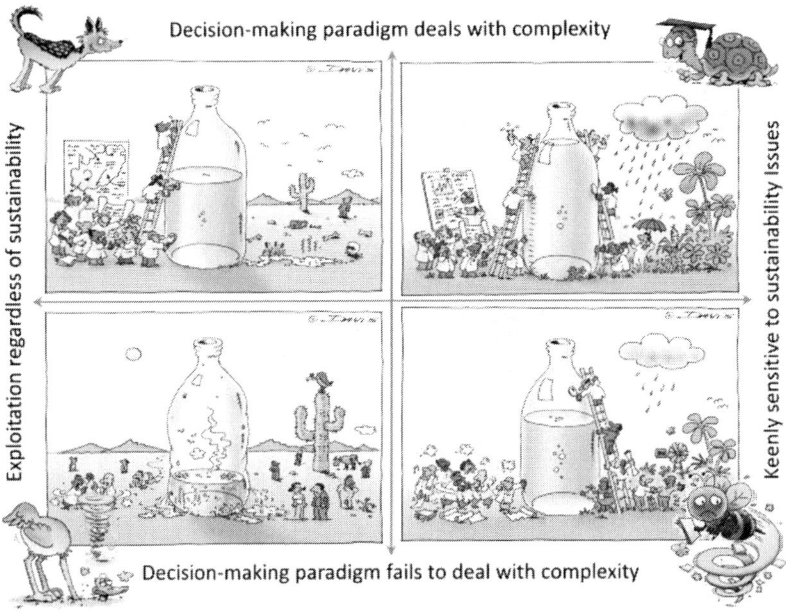

Figure 9. The Water sector institutional landscape by 2025 scenarios [**38**]

2.3.4.1. Dissemination and Impact

The scenarios were printed by the WRC in the form of a colourful booklet and subsequently have been disseminated to some stakeholders. The scenario document and technical report documenting the scenario development process are also available online. While much more could have be done in terms of dissemination, this was not a component required by the project's funders and was therefore not planned into the project process from the start of the project. Therefore no funding was available to carry out this important part of the scenario development process. Nonetheless, these scenarios have the potential of feeding into the decision-making processes of water resources managers and decision-makers, but could also potentially empower a range of other role players in the water sector to engage in participative governance [23].

By studying the dissemination, impact and lessons learned from the South African scenarios discussed above, along with other literature related to the impact of scientific research and the science-policy interface, it is possible to distil some lessons and challenges relating to impact and how to more effectively produce and disseminate impactful scenario products. A discussion on the impact of scenarios in general and reflections on such impact follows below.

2.4. The Impact Of Scenarios

The previous section explored a number of South African scenarios in terms of their contents and impact. In terms of impact, Chantell Illbury and Clem Sunter refer to the "Wack" test, based on the ideas of Pierre Wack, a key scenarios planner in the 1970s and 1980s. According to this test, scenarios are not deemed important because of their prediction capability. What is important is their ability to influence the mindsets of decision-makers and to encourage them to act [39].

The issue of scenario impact is in many ways tied to a broader issue often referred to as the science-policy or science-end user interface. This issue essentially speaks to the challenge of getting knowledge that is produced by scientific or expert teams to be used in the public domain. This discourse recognises that there should be a close relationship between science or research products and their end-users, which could include government, policy-makers, businesses and communities. In reality, however, this relationship is not always an effective one, resulting in research often (or mostly) having minimal impact on policy and practice. The science-policy interface discourse explores why this happens in order to try to advise scientists and end-users about how to more effectively incorporate research into practice [40, 41, 42].

In terms of scenarios there tend to be two major opportunities for impact. The first is an impact on the participants who are part of the scenario development process. This is referred to as 'communication*for* scenarios' [43]. Similarly this opportunity for impact can be referred to as first order influence. First order influence refers to participants in the scenario development process undergoing personal changes in their thinking and behaviour. They also commit to the process, learn new skills, and build new networks and relationships. Because participants increasingly respect, understand and trust each other, they jointly commit to change [37].

The second is the impact of scenarios on broader society. This can be referred to as 'communication *of*scenarios' [43]. Here a wider group of stakeholders ideally need to be exposed to the scenarios once they are fully developed. As such, at this stage it is important to think about ways to foster appropriate dissemination and use of scenarios. This stage can also be referred to as second and third order influence. Second order influence is closely linked to first order influence. Participants who have been part of the scenario development process go back to their communities and networks and start sharing their new language, thoughts and insights with others. Third order influence is a process of social change, but can be difficult to monitor and study because of the many variable factors that influence every change process [37].

The following sub-sections reflect on the impact of scenario development on the participating team as well as the impact or influence of scenarios on broader society.

2.4.1. Impacts On Participants In The Scenario Building Process

Participants in a scenario development process actively engage and transform the process in the sense that they are asked to share their views, ideas, concerns and experiences in order to generate drivers to develop scenarios or stories from these drivers. It is important to recognise that this kind of individual impact is difficult to quantify and tends to be very subtle [35; 37]. Nonetheless, the kinds of impacts that individuals experience can include:

- Experiencing reframed mental models – By being forced, through the scenario development process, to articulate and share different perspectives and mental models, participants are made to think carefully about their perceptions and often re-think their views when faced with other participants' views and the need to move collectively towards a desired future [44].

- Gaining a broadened network of relationships – Scenario development processes bring together groups of people to have open and constructive conversations. This process fosters a shared understanding, trust and a sense of community [44].

- Regenerating energy, commitment, and action – By clarifying desired futures and building consensus about how different actions will navigate society towards certain scenarios, a sense of regenerated energy and commitment can be achieved. Also, with new commitment in place, new actions can be catalysed [44].

- Taking pride in participation – When interviewed, participants tend to be quite proud of their involvement in scenario development processes. This encourages them to use and share the learning from the scenario development process during other projects and/or engagements [35].

- Creating a common vocabulary, trust and mutual understanding – Through the process of developing scenarios these subtle processes tend to be fostered. This is important as it is through trust and understanding that people are able to work together towards a desired shared future [37].

2.4.2. Facilitating and Forming A Scenario Team

Whilst it is clear that a subtle process of impact and transformation can occur in a scenario team, this does not happen automatically. There are a number of lessons that have been learned through scenario development processes over the years that need to be borne in mind.

Firstly, having a diverse team is important [45]. The team should come from different age, race and gender brackets as well as a wide range of ideological spectrums [35]. This diversity is important because the more diverse the team is, the more diverse the driver inputs will be and as such the richer and more accurate the scenario development process will be. Also, an inclusive rather than exclusive scenario development process lends legitimacy to the process [46].

Secondly, embracing transdisciplinarity in any scenario development process is important. This implies that in order for scenarios to have the impact they need, they should be produced by a team made up of multiple different actors from government, civil society, communities, and research institutions. This will help the team to take into account different types of knowledge that different actors have (such as technical, traditional, experiential, cultural, and political knowledge). In so doing the inherent complexity in future planning processes will be reflected [47].

Finally, working with a diverse team with different knowledge, experience and viewpoints is not always easy. Conflict can arise when participants with different viewpoints are made to work together. Also, meaningfully incorporating feedback from diverse sets of stakeholders tends to be a highly time consuming process. Given these and other challenges that can arise, the importance of having a skilled, sensitive and insightful facilitator cannot be underestimated. Such a facilitator needs to be able to manage strong individuals who dominate conversation with their own agendas, and needs to be able to encourage everyone to express their opinions during the scenario development process [35].

2.4.3. Impacts On Broader Society

The impacts of scenarios on broader society are harder to ascertain and measure than the impacts of scenario development on the scenario team itself. This is because there are no measurable criteria for quantifying the impact that scenario products have on society, be they in written or oral form. Also the outcome of scenario development processes can never be attributed to a single factor. Scenario development processes typically deal with broad developmental issues making the range of issues and actors that they try to affect diverse. Scenario development processes also happen within the context of a range of related social activities, such as developments in policy, civil society events and public debates. For example, in the case of developing the South African scenarios of the 1980s and 1990s, there were multiple social forums, political parties, and government groups working on transforming the country. These scenarios and their related processes were just one input amongst many others that were part of the broad transition process. Similarly, the Water Sector Institutional Landscape by 2025 scenarios exist alongside scenarios established by the National Water Resources Strategy, the various government departments that do strategic planning and forecasts in relation to water, and the host of grassroots organisations that work on managing water sustainably for the future. Any impact or change in the water sector must then be attributed to a whole range of interlocking factors rather than just one set of scenarios.

2.4.4. Facilitating The Effective Dissemination Of Scenario Products To Society

In order for scenarios to have influence in the broader public space a number of key lessons are important. Firstly, a broad and extensive communication process is a key requirement and should be planned and budgeted for from the beginning of the project [42]. It is important that such a process targets multiple different actors in society, and takes place at many levels of scale (local, provincial, national) [23, 41, 42] in order to engage society and attempt to create a better future [37]. Non-government actors are an important target audience because they are critical in terms of instigating social debate, bringing about grassroots changes and challenging authorities to improve their performance [42].

In government, actors need to be aware of scenario products and how they can make use of them [23]. With regard to the South African water sector in particular, there seems to be a need to enable officials from DWA to apply the outcome of scenarios thinking and processes in their strategic decision-making aimed at mapping out the future of the water sector. A possible way of enabling experts and government officials to think imaginatively and creatively about the future, given their considerable daily workload and challenges, would be to involve scenario experts as facilitators for strategic planning sessions. Such sessions should ideally take the form of one or two day workshops in order to remove government officials from their immediate working environment and enable them to apply their minds to thinking creatively and focusing exclusively on the planning task at hand [23]. When engaging with government departments, it is important to be sensitive to and aware of different issues inherent in the government hierarchy. Non-political, technical experts tend to have a good knowledge of technical issues, but it is also important to target more senior political actors as they tend to have more decision-making power and can therefore implement changes and ideas brought about through the scenario development process more effectively [42].

It has been argued that regardless of which actor is being focused on, there are three key points to bear in mind in terms of targeting actors with information. A clear plan of action needs to be laid out and followed up on. The information needs to be shared in a manner that is non-threatening, interactive and flexible. Scenarios can be disseminated by tapping into existing networks and events such as management meetings, seminars and the media [42].

In addition, the way that scenarios are packaged and communicated is important [42]. There is a whole host of ways that information can be packaged and disseminated. There can be face-to-face dissemination [23], where scenarios are verbally presented at workshops, conferences, public gatherings, business breakfasts, and corporate events. Style of presentation is crucial in this regard. The presentations need to be simple, clear and memorable. The presenter needs to be engaging and open to feedback from the audience [35]. Radio or television documentaries can also be utilised to disseminate scenario ideas and generate public debate [36, 40, 42].

Another option is to publish the scenarios in a written format. A range of media can be used. The scenarios can be published in books, illustrated pamphlets [23] and newspapers [48]. Cartoon artists can be brought on board to illustrate the scenarios. Magazines and web pages can also be targeted. Written documentation about scenarios has proved to be a successful model. For example, Sunter and Illbury's 'The Mind of the Fox: Scenario Planning in Action' [49] is popular reading material and widely distributed.

Finally, it is crucial that the scenario products are seen as legitimate from the start of the scenario development process. They need to have buy-in from influential people involved in the issue that the scenarios explore. This legitimacy is generated by ensuring that the facilitators of the process as well as the scenario team are respected. Although a range of actors must be included in the scenario development process, and must be targeted in the dissemination process, it remains important to include high level and well-connected people in the team as it is often these individuals who will provide the 'insider' links for scenarios to be heard and disseminated through channels of influence [42]. If these strategic individuals cannot be made part of the team, they need to be made aware of and kept informed about the scenario development process to secure their interest and support [35].

Dissemination is not without its challenges. It is challenging to disseminate in a way that suits and reaches a diverse audience with different languages, levels of education, varied professional backgrounds and cultures. Another challenge of ensuring the uptake of scenarios (and research in general) is that dissemination is often not part of the project planning process, and as a result funding often runs out before scenario uptake and use can be promoted [48]. Also, depending on how it is done, the dissemination of scenarios can be very expensive [35].

2.4.5. General Reflections On Impact

Over and above the specific processes linked to the impact of scenarios on the scenario team and broader society there are some general points that are important to bear in mind when planning for impact in relation to scenario products.

Firstly, when starting the scenario development process, it is important to be clear about the purpose of the process one is undertaking and designing it accordingly. What are the intended outcomes of the process? Who is the process meant to influence and what product(s) will be necessary for this to happen [35]? Essentially scenarios need to fill a strategic gap or opportunity in society [50].

Secondly, questions also need to be asked about the timing of the scenario development process. Is there likely to be sufficient recognition among the intended target audience(s) that the problem being addressed is important and that the process is therefore potentially beneficial? Is the political environment such that intended target audience(s) will be responsive to fresh, unorthodox thinking [35,42]?

Thirdly, attention also needs to be paid to the legitimacy of those financing and promoting the process, and the credibility of the project team developing the scenarios in the eyes of both the sponsors and the target audience(s) [35, 40, 42].

3. CONCLUSIONS

In conclusion, it seems that since the initial High Road/Low Road scenarios were developed, scenario development has taken root in South Africa, with several follow up scenarios having been developed since [28]. This development suggests that South African decision-makers must deem scenario development to be of considerable importance and utility, as it is often government or government-related institutions that develop or commission new sets of scenarios. These subsequent scenarios seem to mirror their predecessors with their snappy titles and straightforward structure and certainly have the potential to inspire decision-makers with regard to their planning activities [28].

Based on the discussion and reflections above, scenario development should involve a focus on dissemination and impact from the onset of the scenario process. Impact can happen at the level of participants in the scenario development process as they are exposed to new ideas and start adopting a new way of thinking about current issues of importance. These ideas have the potential to slowly infiltrate the networks of these participants and to also influence their thinking. At the same time, it is important to have a strong dissemination process in order to reach as many people as possible beyond the project team. The High Road/Low Road presentations are an example of a highly effective dissemination process made possible by an engaging speaker and interesting topic that was clearly and simply brought across to a wide range of audiences. Another key method of dissemination is to raise awareness about where the scenarios can be found and to make it easy for people to access them. The open access route followed by the Dinokeng scenario team is a good example of a scenario document that is easily available on a website, accompanied by much useful background information. It is this dissemination phase that has been lacking in the Water Sector Institutional Landscape by 2025 scenarios, and a follow up process is needed to plan how more people could be made aware of these scenarios and their usefulness to decision-making and planning in the South African water sector.

It is also important to keep in mind that scenarios are likely to have a higher impact if they are developed with the intention of identifying or solving particular problems [51]. If there is an intended target audience with particular information needs from the beginning of the scenario process, the scenario team will be able to keep this in mind when developing the scenarios. This will also ensure more effective uptake of the scenarios as pre-defined end-users exist. In the water sector, for instance, it could be effective for decision-makers who are grappling with a particular issue to solicit scenario inputs to aid them in making decisions regarding that issue.

In terms of future research, three areas come to mind based on what has been discussed in this chapter:

Firstly, a large scale study (mostly comprising of interviews) is needed to understand in greater detail the impact of scenarios on scenario participants, society and government planning processes [28]. Much of what has been argued to date in terms of the impact of scenarios has been on the basis of inference and assumptions. It would be interesting, though admittedly also very difficult, to try to substantiate views around the impact of scenarios with empirical evidence.

Secondly, it would be important to study how a scenario team would know that the timing is right to come up with and disseminate a new set of scenarios. It is reasonably easy to see that scenarios would have been important for particular moments in history, for example the political transition in South Africa, but it is considerably more difficult to determine when there may be an ideal window of opportunity in future in which scenarios may make an impact. It may also be important to determine which factors other than and in support of ideal timing would be important for scenarios to achieve impact.

Thirdly, building on this chapter, it would be important to determine how best to ensure that scenarios can become more useful and practical to policy-makers and other end-users. How can scenario teams ensure that end-users know how best they may use scenarios in order to influence their future planning? The issue of providing navigation to and between different scenarios and future outcomes is important in this regard.

Clearly, scenario development is a useful process to help decision-makers cope with and plan amidst uncertainty. Particularly in the context of the South African water sector, it is important to recognise that uncertainty is deepening in many ways given the impending presence of multiple stressors such as climate change, basin closure, growing populations, migration movements and a growing economy. These stressors, along with the institutional fluctuations and changes within the water sector itself, make it increasingly important for decision-makers to work with scenarios to help them to plan sensibly and creatively despite uncertainty. However, in order for scenarios to be useful it is important to plan for and carefully think about how to maximise their impact.

ACKNOWLEDGEMENTS

The authors would like to acknowledge the CSIR's librarian, Engela van Heerden, for her excellent work in sourcing a large amount of relevant literature that contributed to this chapter. They would also like to acknowledge Wilma Strydom for her valuable review comments on this chapter. Finally, the authors would like to acknowledge the Water Research Commission (WRC), the organisation that funded the development of the Water Sector Institutional Landscape by 2025 scenarios. The learning that the project team gained from this project was a key input into this chapter.

REFERENCES

1. Bogardi JJ, Dudgeon D, Lawford R, Flinkerbusch E, Meyn A, Pahl-Wostl C, Vielhauuer K and Vörösmarty C. Water Security for a Planet Under Pressure: Interconnected Challenges of a Changing World Call for Sustainable Solutions. Current Opinion in Environmental Sustainability 2012; 4 35–43.

2. Slaughter RA.Welcome to the Anthropocene. Futures 2012; 44 119–126.

3. WWAP (World Water Assessment Programme). The United Nations World Water Development Report 4: Managing Water under Uncertainty and Risk. Paris: UNESCO; 2012.

4. UN (United Nations). Declaration of the United Nations Conference on the Human Environment. Stockholm, 5-16 June 1972.

5. Brundtland GH. Address by Mrs Gro Harlem Brundtland, Chairman at the Closing Ceremony of the Eighth and Final Meeting of the World Commission on Environment and Development 27 February 1987. Tokyo, Japan.

6. WMO (World Meteorological Organisation). International Conference on Water and the Environment: Development Issues for the 21st Century. Geneva: ICWE Secretariat; 1992.

7. UN (United Nations). Agenda 21. United Nations Conference on Environment and Development Rio de Janeiro, Brazil, 3 to 14 June 1992.

8. Spangenberg JH, Pfahl S, Deller K. Towards Indicators for Institutional Sustainability: Lessons From an Analysis of Agenda 21. Ecological Indicators 2002; 2 61–77.

9. UN (United Nations). World Development Report, 1980. Washington, D.C.: The World Bank; 1980.

10. UN (United Nations). World Development Report, 1985. New York: Oxford University Press; 1985.

11. UN (United Nations). World Development Report, 1990: Poverty. New York: Oxford University Press; 1985.

12. UN (United Nations). World Development Report, 1995: Workers in an Integrating World. New York: Oxford University Press; 1995.

13. World Bank. Indicators. http://data.worldbank.org/indicator. (accessed 30 July 2012).

14. Bates C, Kundzewicz ZW, Wu S and Palutikof JP., editors. Climate Change and Water. Technical Paper of the Intergovernmental Panel on Climate Change. Geneva: IPCC Secretariat; 2008.

15. Davis CL. Climate Risk and Vulnerability: A Handbook for Southern Africa. Pretoria: Council for Scientific and Industrial Research; 2011.

16. Tewari DD. A Detailed Analysis of the Evolution of Water Rights in South Africa: An Account of Three and a Half Centuries from 1652 AD to the Present. Water SA 2009; 35(5) 693-710.

17. Gleick PH. The World's Water, 1998-1999. Washington DC: Island Press; 1998.

18. DWAF (Department of Water Affairs and Forestry). National Water Act of South Africa. Act No. 36 of 1998. Pretoria: Department of Water Affairs and Forestry; 1998.

19. DWAF (Department of Water Affairs and Forestry). National Water Resources Strategy – First Edition. Pretoria: Department of Water Affairs and Forestry; 2004.

20. DWA (Department of Water Affairs). Annual Report of the Department of Water Affairs Vote 37: 1 April 2010 To 31 March 2011. http://www.pmg.org.za/minutes/30. (accessed 6 August 2012).

21. Emerson JW, Hsu A, Levy MA, de Sherbinin A, Mara V, Esty DC, Jaiteh M. Environmental Performance Index and Pilot Trend Environmental Performance Index. New Haven: Yale Center for Environmental Law and Policy; 2012.

22. Green GC, Day JA, Mitchell SA, Palmer C, Laker MC, Buckley CA. Water Research Commission 40-Year Celebration Conference - Syntheses of Themed Sessions. Water SA 2011; 37(5) 609-618.

23. Claassen M, Funke N, Nienaber S. 2011. The Water Sector Institutional Landscape by 2025 Technical Report. WRC Report No. 1841/1/11. Pretoria: Water Research Commission; 2011.

24. Mietzner D, Reger G. EU-US Seminar: New Technology Foresight, Forecasting and Assessment Methods. Seville; 13-14 May 2004.

25. Landeta J. Current Validity of the Delphi Method in Social Sciences. Technological Forecasting and Social Change 2006; 73 467-482.

26. Daum J. How Scenario Planning Can Significantly Reduce Strategic Risks and Boost Value in the Innovation Chain. The New Economy Analyst Report 8 September 2001.

27. Sunter C. The World and South Africa in the 1990s. Cape Town: Human Rousseau Tafelberg; 1987.

28. Galer G. Scenarios of Change in South Africa. The Round Table 2004; 93(375) 369-383.

29. DACST (Department of Arts, Culture, Science and Technology). South African National Research and Technology Foresight Project. Pretoria: DACST; 1999.

30. Dinokeng team. 3 Futures for South Africa. www.dinokengscenarios.co.za (accessed 10 August 2012).

31. (WBCSD) World Business Council for Sustainable Development. Business in the World of Water: WBCSD Water Scenarios to 2025. www.wbcsd.org/web/H2OScenarios.htm (accessed 30 July 2012).

32. GRA (Global Research Alliance) Science and Technology-Based Water Scenarios for Sub-Saharan Africa. www.research-alliance.net. (accessed 6 August 2012).

33. Peterson GD, Cumming GS, Carpenter SR. Scenario Planning: A Tool for Conservation in an Uncertain World. Conservation Biology 2003; 17(2) 358-366.

34. Finlev T. Future Peace: Breaking Cycles of Violence through Futures Thinking. Journal of Futures Studies 2012; 16(3) 47-62.

35. Segal N. Breaking the Mould: The Role of Scenarios in Shaping South Africa's Future. Stellenbosch: SUN Press; 2007.

36. Kahane A. The Mont Fleur Scenarios: What Will South Africe Be Like in the Year 2002? Deeper News 1992, 7(1) 1-22.

37. Maddison S, Cronin D, Williams S, Coggan R. Democratic Dialogue: Finding the Right Model for Australia. Indigenous Policy and Research Unit, Discussion Paper No.1. Sydney: University of New South Wales; 2009.

38. Claassen M, Funke N, Nienaber S. The Water Sector Institutional Landscape by 2025. WRC Report No. TT 514/11. Pretoria: Water Research Commission; 2011.

39. Karuri-Sebina G, Rosenzweig L. A Case Study on Localising Foresight in South Africa: Using Foresight in the Context of Local Government Participatory Planning. Foresight 2007, 14(1) 26-40.

40. Strydom WF, Funke N, Nienaber S, Nortje K, Steyn M. Evidence-based Policy-making: A Review. South African Journal of Science 2010; 106(5/6) 249 – 334.

41. Funke N, Nienaber S, Henwood R. Scientists as Lobbyists? How Science Can Make its Voice Heard in the South Africa Policy-Making Arena. International Journal of Public Affairs 2011; 10(102) 421 – 443.

42. Funke N, Nienaber S. Promoting Uptake and Use of Conservation Science in South Africa by Government. Water SA 2012; 38(1) 105-114.

43. Lebel L, Thongbai P, Kok K, Agard JBR, Bennett EM, Biggs R, Ferreira M, Filer C, Gokhale Y, Mala W, Rumsey C, Velarde SJ, Zurek M, Blanco H, Lynam T, Tianxiang Y. Sub-global Scenarios. In: Capistrano D, Lee M, Raudsepp-Hearne C, Samper C. (eds.) Ecosystems and Human Wellbeing: Multi-scale Assessments. Findings of the Subglobal Assessments Working Group of the Millennium Ecosystem Assessment. Washington D.C.: Island Press; 2005. p229-259.

44. Ringland G. Scenario Planning, 2nd edition. England: John Wiley & Sons; 2006.

45. Audouin M, Preiser R, Nienaber S, Downsborough L, Lanz J, Mavengahama S. Exploring the Logic of Complexity for Researching Social-Ecological Systems. Ecology and Society In Press.

46. Ogilvy JA. Creating Better Futures: Scenario Planning as a Tool for a Better Tomorrow. New York: Oxford University Press; 2002.

47. Jacobs I, Nienaber S. Water Without Borders: Transboundary Water Governance and the Role of the 'Transdisciplinary Individual' in Southern Africa. Water SA (WRC 40 Year Celebration Special Edition 2011; 37(5) 665 – 678.

48. Communication with Clem Sunter. 26 July 2012. Pretoria.

49. Illbury C, Sunter C. The Mind of a Fox. Cape Town: Human & Rousseau/Tafelberg; 2001.

50. Johnson G, Scholes K, Whittington R. Exploring Corporate Strategy: Text and Cases. New Jersey: Prentice Hall, Financial Times Press; 2008.

51. Bohenksy E, Reyers B, Van Jaarsveld AS. Future Ecosystem Services in a Southern African River Basin: a Scenario Planning Approach to Uncertainty. Conservation Biology 2006; 20(4) 1051-1061.

Chapter 5

Water Quality Aspects of River Okpoka in Relation to Watershed Activities

Nevoh G. O. [1], Akhionbare S. M. O. [2], Uzoma H. C. [3]

[1]Petroleum Analysis Laboratory, Petroleum Training Institute Effurun Warri, Nigeria

[2]Department of Environmental Technology, Federal University of Technology Owerri, Nigeria

[3]Department of Environmental Technology, Federal University of Technology Owerri, Owerri, Nigeria

ABSTRACT

The quality of River Okpoka in south-south Nigeria was investigated by a field survey during the dry season. Water and sediment samples were collected from three sites (capturing the activities in an Abattoir, a market and residential community) along the River course. Three fish species (Tilapia sparrmanii, Mugilcephalus and Clariasgariepinus) were also collected for heavy metal analysis. Samples were analysed for microbiological and physicochemical parameters using standard methods. Heavy metals (Ni, Fe, Pb, Hg, Zn, Cr and Cd) in water, sediment and fish samples were carried out using Atomic Absorption Spectrophotometer (AAS). Descriptive statistics and Pearson correlation were used to analyze data at $P<0.05$ and 0.01. Generally, the range of value for Chloride (6780.50–7188.75mg/L), Sulphate (534.67–633.33mg/L), Conductivity

(21400.00–24900.00μS/cm), TDS (10720.00–12450.00mg/L), BOD (4.58–4.96mg/L), Total hardness (121.25–117.50mg/L) and coliform counts (1.40x101–3.6x101cfu/m) across the sites shows elevated levels. Cd (3.98–4.36mg/L), Ni (6.75–15.50mg/L) & Fe (0.60–2.20mg/l) also recorded elevated levels with relative abundance in the order Ni>Cd>Fe>Pb>Zn>Cr>Hg. Heavy metal values from sediment recorded low concentrations, with Fe (60.60–73.40mg/kg), Zn (2.58–5.93mg/kg) and Cr (3.00–9.50mg/kg) recording significant values in relative order of Fe>Cr>Zn>Pb>Ni>Cd>Hg. Cr (2.50–5.00mg/kg) and Ni (1.40–6.60mg/kg) in fish species were at elevated levels, with relative order of abundance in the order Cr>Fe>Ni>Zn>Cd>Pb>Hg. Significant correlations at $p<0.01$ were recorded for Conductivity with TDS (+1.00**), and pH &Alkalinity (–1.00**). At $p<0.05$, Cd and Coliform count (+1.00*), Ni and Sulphate (–0.997*). Several significant correlation at $p<0.01$and $P<0.05$ were recorded amongst heavy metal in fish samples and those in water; Fe and Pb (–0.946**), Cr and Cd (–0.870*), Zn and Cd (–0.869), Ni and Cd (+0.867*). Water quality index value of 22 indicated that River Okpoka is almost always threatened or impaired; conditions usually depart from natural or desirable levels. The heavy metal species and some physical and chemical parameters in the River Okpoka were to a large extent above the WHO limits; and thereby pose a health risk to rural communities who rely heavily on the river primarily as their source of domestic water. The study revealed a need for continuous pollution monitoring and management program of surface water in River state and Nigeria as a whole.

KEYWORDS

Sediments, River state, Anthropogenic, Relative abundance, Heavy metals

1. INTRODUCTION

World population growth continues to escalate rapidly, accompanied by rapid urbanization, industrialization and increased demand for food (Chamie, 2004). These factors have consequently led to over exploitation and pollution of already stressed freshwater resources leading to deterioration in water quality, a reduction in water quantity and degradation of freshwater biodiversity habitats (Jewitt, 2002; Hassan et al., 2005). Industrialization over the past half century has led to huge increases in the discharge of toxic chemicals into fresh water bodies, some of whose toxicity are partly or totally unknown (Hassan et al., 2005). In addition, many of these chemicals are persistent and could transform into by-products that may have adverse effects on water resources (Hassan et al., 2005). As a result, the ability of freshwater ecosystems to provide clean and reliable sources of water, maintain the natural water cycle and the biological food web as well as provision of food and recycling of nutrients have been severely impaired (Hassan et al., 2005). These have limited the amount of useable water available for biodiversity, for further economic and social developments as well as natural ecosystems functioning (Jewitt, 2002; Hassan et al., 2005). The occurrence of heavy metals in aquatic ecosystems in excess of natural background loads has become a problem of increasing concern due to their toxicity, persistence and bioaccumulation problems and their tendency to accumulate in vital human organs (Obasohan et al., 2006; Olayinka, 2004; Greaney, 2005; Akoto et al., 2008; Defew et al., 2004; Tam and Wong, 2000). The presence of trace metals in aquatic systems originates from the natural interaction between water, sediment and the atmosphere. The physical and chemical characteristics of water are important parameters as they may directly or indirectly affect its quality and consequently its suitability for the distribution and production of fish and other aquatic animals (Swingle, 1969; Moses, 1983). The Okpoka River is one of the most important river systems is the Niger delta. It is exceptionally prone to impacts, because of the numerous human activities that are carried out: fishing, boat construction; welding, sand mining, dredging, mangrove cutting, and engine boat transportation, etc which may be potential sources of pollution to the environment. Besides, there are a lot of municipal waste discharges such as human waste, waste from abattoir, refuse dumps, sewage/septic waste, etc into the River water. Also the presence of several oil servicing companies whom engages in effluent discharges contribute to compound the woes of River Okpokais to investigate the levels of water variables and selected heavy metals species and establish the quality of water in River Okpoka, River state Nigeria.

2. METHODOLOGY

2.1 Study Area

The study area is located in the Port Harcourt Local Government area of River state of Nigeria, within the mangrove forest vegetation belt of Nigeria. River Okpoka originates from Etche and lies between Longitudes 7°

03' and 7°05' East and Latitudes 4°06' and 4°24' and it is about 6 kilometers long and continues to Abulomatown of Rivers State where it finally joins the Bonny River before emptying into the Atlantic ocean. Port Harcourt features a tropical mansoon climate with lengthy and heavy rainy season and very short dry season. Only the months of December and January truly qualifies as dry season months in the city. The harmattan which climatically influences many cities in West Africa is less pronounced in Port Harcourt. Its heaviest precipitation occurs during September with an average of 370mm of rainfall. December on average is the driest month of the year with an average rainfall of 20mm. Temperatures throughout the year is relatively constant showing little variations. Average temperatures are typically between 25°C–28°C (Abu and Egeronu, 2004). Anthropogenic activities in the study area include agricultural, commercial, industrial, automobile/boat and ship building, Proliferation of both human and solid waste dumps.

2.2 Sampling

Three locations separated by at least 1km from each other along the river course were established using GPS device and labelled 3 (Upsteam), 1, 2 (downstream). Water and sediment samples were collected at each location during the dry season in the month of March.

Sample Analysis: River water were analysed for physicochemical parameters and heavy metal, sediment and fish samples were analyzed for heavy metal concentrations according to standard methods for the parameters (APHA, 2012) and Atomic Absorption Spectrophotometer (AAS) for heavy metals. Possible metal-metal, metal-pH relationships were investigated using the Pearson correlation coefficient to determine significant levels (95% confidence). The procedure below was used to estimate the Water Quality index (WQI) of the water body. The procedure employed the method of Horton (1965) as modified by Ramakrishnalah (2009).

Ø According to the relative importance of the chemical parameters in the overall quality of water for drinking purposes specific weights are assigned to them. The most significant parameters have a weight of 5 and the least significant a weight of 1. Nitrate playing a prominent role in ground water quality for drinking purposes (Ramakrishnalah et al., 2009), thus maximum weight of 5 has been assigned to this parameter whereas the magnesium is not harmful to ground water quality for drinking purposes and the weight of 1 is given to the magnesium (table 1).

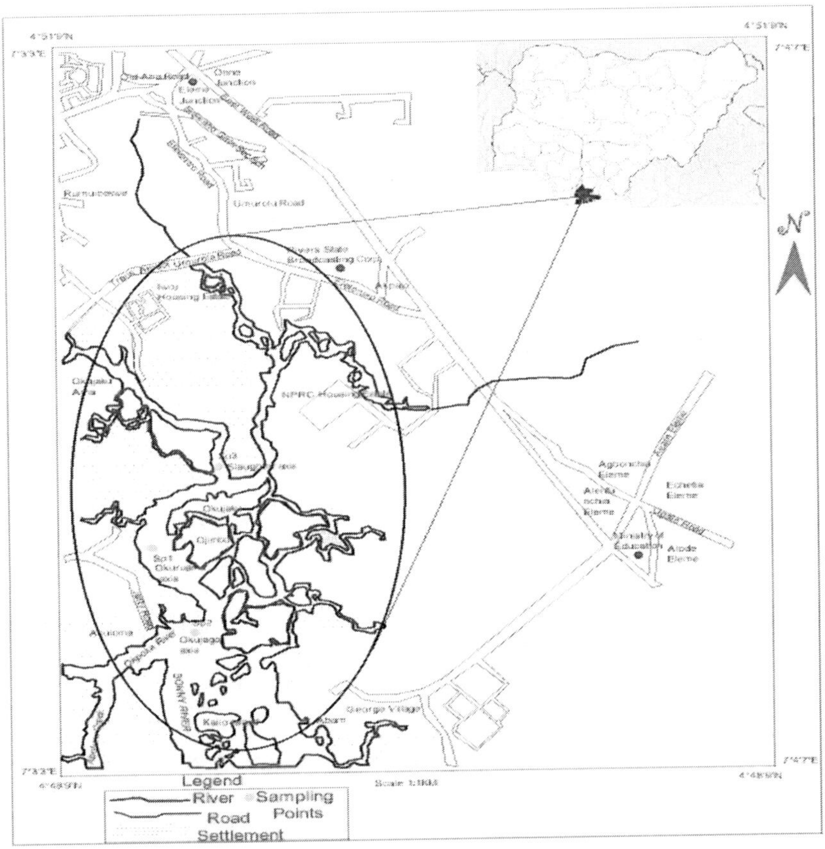

Figure 1. Map of study area showing sampling points

Ø Compute the relative weight (Wi) by using the following equation:

$$W_i = \frac{w_i}{\sum^{n} w_i} \quad (1)$$

Where, Wi is the relative weight, wi is the weight of each parameter and n is the number of parameters.

Ø Calculate a quality rating scale (qi) for each parameter which is assigned by dividing its concentration in each water sample by its respective standard.

According to the guidelines laid down in the BIS and the result multiplied by 100.

$$qi = (Ci / Si) * 100 \qquad (2)$$

Table 1. Water quality classification based on WQI value

WQI Value	Class	Water Quality
<50	I	Excellent
50-100	II	Good water
100-200	III	Poor water
200-300	IV	Very poor water
>300	V	Unsuitable water

Source: Al-Hadithi (2012).

Where qi is the quality rating, Ci is the concentration of each chemical parameter in each water sample in mg/L, and Si is the Indian drinking water standard for each chemical parameter in mg/L according to the guidelines of the BIS 10500, 1991.

Ø Determined SIi for each chemical parameter which has been calculated from the following equation

$$SIi = Wi \, .qi \qquad (3)$$

Ø Finally, compute WQI for the sample by using the following formula:

$$WQI = \Sigma \, SIi \qquad (4)$$

The computed WQI values are classified into five types, "excellent water" to "water, unsuitable for drinking" (Table 2).

3. RESULTS AND DISCUSSION

Figures 1 - 3 show the variation of concentrations of water variable in River Okpoka. Also figures 4 and 5 show the variations of heavy metals in sediment and fish species Table 1–6 represents: GPS Coordinates, Descriptive statistics of water variables, Descriptive statistics of microbiological variable, Descriptive statistics of heavy metals in water, Descriptive statistics of heavy metals in sediment and Descriptive statistics of heavy metals in fish during the study periods of March, 2014.

Table 2. GPS coordinates of sampling locations within the River Okpoka in Port Harcourt

			Sampling points
	1	2	3
Northing	4° 48′48.23″	4° 48′46.69″	4°48′54.14″
Easting	7°2′48.43″	7° 3′ 3.74″	7°2′ 46.99″

3.1 pH

The spatial distribution of pH observed in this study is characteristic of a tidal brackish water environment as noted by the International Joint Commission, (1977) and Ajao and Fagade, (2002). The pH value of this study ranged 7.10–7.40 (Table 3). The higher pH values at the upstream (SP 3) than at the downstream (SP 1 & 2) (fig 1). may have resulted from decaying of the domestic and industrial waste litter in the upstream area contributing to the basic nature of the water. However, pH values recorded in this study were well within the preferred pH of 6.5 to 9.0 recommended for optimal fish production (Boyd and Lichktopller, 1979). The pH values observed are in agreement with results of previous studies conducted by Dublin-Green (1990) in Bonny River, where the highest pH values were recorded in the dry season and lower values of pH in the late rainy season. Similar trend was reported by Ekeh and Sikoki (2003) in the New Calabar River and also by Ansa (2005) in Andoni of the Niger Delta area.

Table 3. Descriptive statistics of physico-chemical variables in water of River Okpoka, Port Harcourt

Parameters	SP1	SP2	SP3	Minimum	Maximum	Range
pH	7.10	7.36	7.40	7.10	7.40	0.30
TotalAlkalinity (mg/l)	126.88	98.82	93.94	92.94	126.88	32.00
Chloride (mg/l)	6780.50	7011.25	7188.75	6780.50	7188.75	408.25
TotalHardness (mg/l)	117.50	150.00	121.25	117.50	150.00	32.50
Sulphate (mg/l)	633.33	566.67	534.67	534.67	633.33	98.66
Nitrate (mg/l)	16.10	10.70	13.30	10.70	16.10	5.40
Conductivity (µs/cm)	21400.00	24800.00	24900.00	21400.00	24900.00	3500.00
TDS (mg/l)	10720.00	12410.00	12450.00	10720.00	12450.00	1730.00
Salinity (psu)	12.82	15.09	14.92	12.82	15.09	2.27
BOD(mg/l)	4.96	4.58	4.64	4.58	4.96	0.38

SP=Sampling Point.

Table 4. Descriptive statistics of Microbiological variables in water of River Okpoka, Port Harcourt

Parameter	SP1	SP2	SP3	Minimum	Maximum	Range
Coliformcount (cfu/m)	18.00	14.00	36.00	14.00	36.00	22.00

SP=Sampling Point,

Table 5. Descriptive statistics of Heavy metals variables in water of River Okpoka, Port Harcourt

Parameters	SP1	SP2	SP3	Minimum	Maximum	Range
Fe (ppm)	2.20	0.60	1.60	0.60	2.20	1.60
Pb (ppm)	<0.001	<0.001	<0.001	<0.001	<0.001	<0.001
Ni (ppm)	6.75	13.25	15.50	6.75	15.50	8.75
Zn (ppm)	<0.001	<0.001	<0.001	<0.001	<0.001	<0.001
Cd (ppm)	4.06	3.98	4.36	3.98	4.36	0.38
Cr (ppm)	<0.001	<0.001	<0.001	<0.001	<0.001	<0.001
Hg (ppm)	<0.001	<0.001	<0.001	<0.001	<0.001	< 0.001

SP=Sampling Point

Table 6. Descriptive statistics of Heavy metals variables in Sediment of River Okpoka, Port Harcourt

Parameters	SP1	SP2	SP3	Minimum	Maximum	Range
Fe (ppm)	60.6	73.40	66.00	60.60	73.40	12.80
Pb (ppm)	1.50	<0.001	<0.001	<0.001	1.50	0.75
Ni (ppm)	<0.001	<0.001	<0.001	<0.001	<0.001	<0.001
Zn (ppm)	5.93	4.37	2.58	2.58	5.93	3.35
Cd (ppm)	<0.001	<0.001	<0.001	<0.001	<0.001	<0.001
Cr (ppm)	3.00	9.50	8.50	3.00	9.50	6.50
Hg (ppm)	<0.001	<0.001	<0.001	<0.001	<0.001	<0.001

SP=Sampling Point

Table 7. Descriptive statistics of Heavy metals in Fish species sampled downstream (SP2)

Parameters	Mullet	Cat	Tilapia
Ni (mg/kg)	1.50	3.00	2.25
Fe (mg/kg)	1.40	2.60	6.60
Pb (mg/kg)	<0.001	<0.001	0.50
Hg (mg/kg)	<0.001	<0.001	<0.001
Zn (mg/kg)	1.07	1.47	1.48
Cr (mg/kg)	4.00	2.50	5.00
Cd (mg/kg)	<0.001	0.12	0.10

3.2 Salinity

The spatial distribution of salinity values (range= 12.82–15.09) (Table 2) showed gradual increase of salinity values from the upstream stations to downstream stations along the creek. This trend could be attributed to effluent water discharges from several industrial establishments, slaughterhouse operations and domestic activities that are prevalent along the upstream area of the creek. Higher salinity value recorded compared favorably the report by Payne, (1976), whom recorded higher salinity values during the dry season. The months of April to October in West Africa usually coincide with the rainy season when high volumes of freshwater are discharged into coastal or estuarine waters that lower or dilute the water. Similarly, Akhionbare & Akhionbare (2005), Akhionbare (2007) reported that rainfall could cause dilution of estuaries and hence cause reduction in salinity.

3.3 BOD

Biochemical Oxygen demand (BOD) exhibited similar values across the sampling stations with higher values recorded at the upstream stations 1 than at the downstream stations (Table 3). The higher BOD (range= 4.58–4.96mg/l) load observed in the months of March could be attributed to increased degradable organic waste load observed during the study as noted by McNeely et al. (1979) and Clerk, (1986). The Biochemical Oxygen Demand values were generally stable in all stations (fig 1) and exhibited no significant difference ($P<0.05$) temporal variations. Based on the observed results in this study, it could be stated that the BOD values may be considered as "doubtful water quality" since they have BOD that is close to 4.0mg/l. This assertion follows the classification of Moore and Moore (1976) who opined that water bodies with BOD concentrations between 1.0 and 2.0mg/l were considered clean 3.0mg/l fairly clean, 5.0m g/l doubtful and 10.0mg/l definitely bad and polluted. BOD concentrations in the creek may therefore serve as a pointer to the level of organic pollution. This agreed with the observation by Braide et al. (2004) in their study on water quality of Miniweja stream in Eastern Niger Delta.

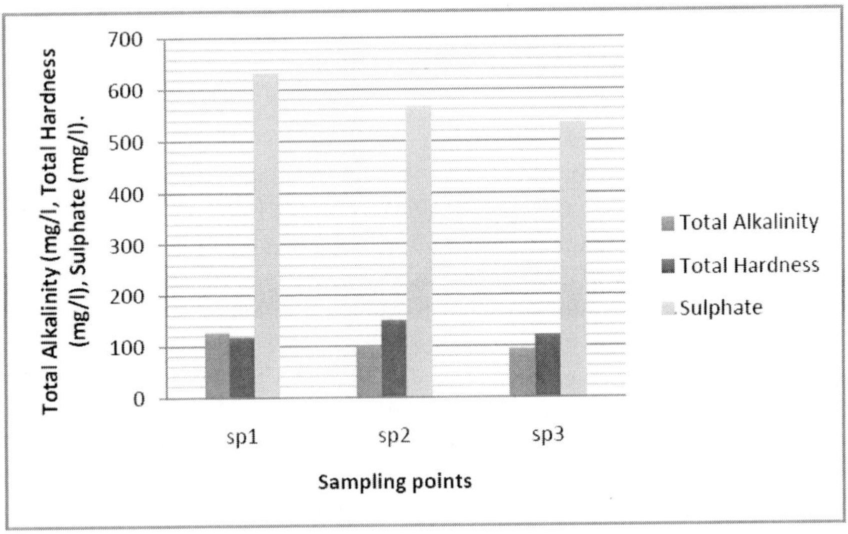

Figure 1a. Spatial variation of water variables across sampling points

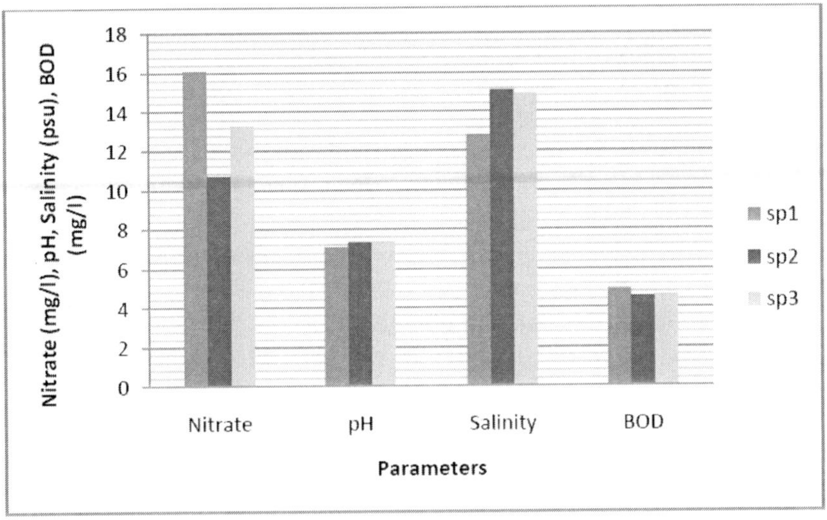

Figure 1b. Spatial variation of water variables across sampling points

3.4 TDS/Sulphate

The results obtained for TDS in this study shows highly elevated values, all exceeding the WHO standards for drinking water (500mg/l) (Table 3). The highest value (12450.00mg/l) was recorded at sampling station 3. This is the area around the slaughter house which receives large volume of discharge from animal slaughter operations, market and waste disposal. Sulphate levels in this study are quite high (534.67–633.33mg/l), increasing downstream (Fig 2). These values are by far higher than the WHO standards for drinking water (250mg/l). There are several sources of sulphate in River water; decaying plant and animal matter may release sulphate into water. Numerous chemical products including ammonium sulphate fertilizers contain sulphate in varying form. The consumption of drinking water containing high amount of magnesium or sodium sulphate may result in intestinal discomfort, diarrhoea and consequently dehydration. (Akhionbare, 2009).

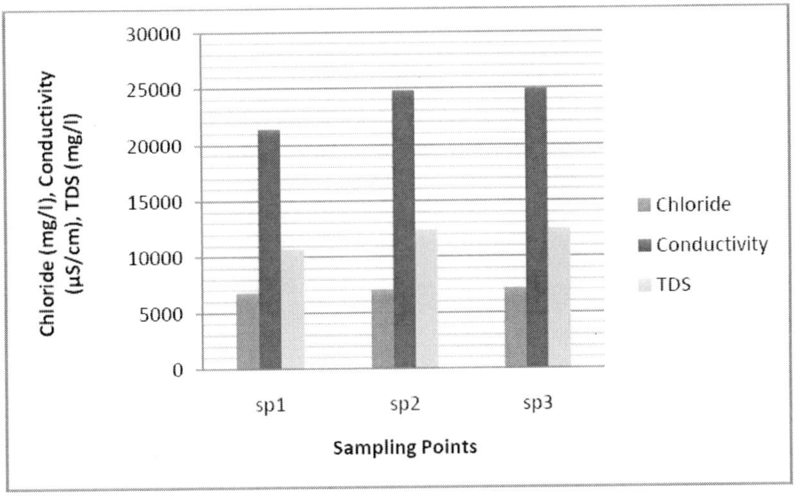

Figure 2. Spatial variation of water variables across sampling points

3.5 Heavy Metals

Heavy metals value obtained in this study from the water sample of River Okpoka showed elevated values (Table 5) for Fe (range= 0.60–2.20mg/l), Ni (6.75–15.50mg/l) and Cd (range= 3.98–4.36mg/l), with values exceeding the WHO standard for drinking water. Pb, Zn, Cr and Hg recorded negligible values (fig 3). Anthropogenic sources of iron in an urbanized environment are largely due to corrosion of iron and steel structure in buildings and vessel (evident in River Okopka with several corroded steel vessel floating), dumpsites and abandon vehicles. The main source of Cd evident within the River Okpoka watershed is from waste water discharge. Ni is a component of some alloy and also used in battery. Indiscriminate disposal of solid waste on water body is a major source of Ni likewise sewage and household waste water as evident within the watershed. Figure 4 depicts the variations of the heavy metals across the stations.

Generally, all the heavy metals analysed in the sediment sample of this study occurred at low levels and this is attributable to high self purification capacity. Rivers are "open" systems with continual renewal of water, and therefore have high capacity for self-purification due to oxygen diffusion at the surface, together with dilution, which enables the water to assimilate polluting discharges. Consequently, in the sediment, all trace metals investigated recorded low concentrations. Fe recorded the highest value

(73.40mg/kg) at sampling station 2, while Ni, Cd and Hg recorded negligible values. Fig 3.

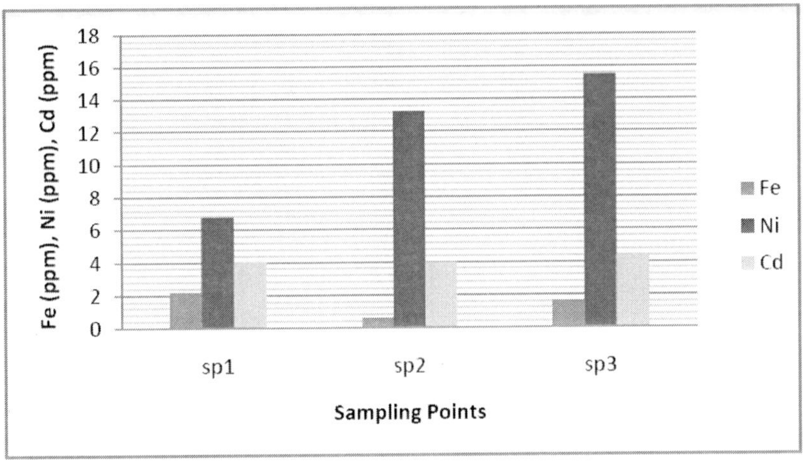

Figure 3. Spatial variation of heavy metals in water sample of River Okpoka across sampling points

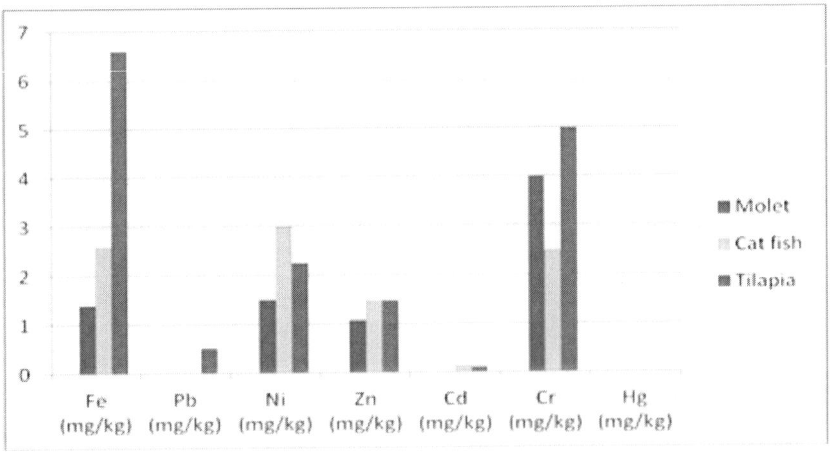

Figure 4. Spatial variation of heavy metals in fish species (Gills and Tissue)

3.6 Fish Species

Iron was the most abundant metal in the studied tissues of all the fish species. In the fishes analyzed, Fe accumulation was the maximum among the 3 species, and its highest value (6.60mg/kg) (fig 4) was observed in Tilapia Tilapiasparrmani. However in other studies the highest accumulation was observed in the organs of gills in the fishes O.niloticus, L. niloticus (Mohammad, 2008) O. mossambicus (Jenny and Avenant, 2006). The highest concentrations observed in this present study were below the high residue concentrations of Fe (34-107 mg/kg) in fish samples (Charbonneau and Nash, 1993). The highest levels of Zn in this study (Table 7) were observed in the tissue of Tilapia Tilapiasparrmanii. The mullet Mugilcephalus and Cat Clariasgariepinus recorded concentrations. However highest accumulation was seen in other organs such as gills of C.punctatus (Vineeta et al., 2007), testis of O. niloticus and L. Niloticus (Charbonneau and Nash, 1993). The permissible limits for Zn set by WHO (1989) is 40 ppm which is higher than the values observed during this study. Zinc was detected in all the fish samples. The concentration of Zn in Tilapia sparrmanii and other species studied were below the NCBP 34.2µg/g. Fish can accumulate zinc from both the surrounding water and from their diet (Eisler, 1993). Although zinc is an essential element, at high concentrations, it can be toxic to fish, cause mortality, growth retardation and reproductive impairment (Sorenson, 1991). Lead is highly toxics to aquatic organisms, especially fish (Rompala et al., 1984). This study recorded low concentration for Pb (<0-001–0.50mg/kg), with significant value observed on Tilapia Tilapiasparrmanii. Other species recorded negligible values. The biological effects of long term concentrations of lead include delayed embryonic development, suppressed reproduction and inhalation of growth, increased mucous formation, neurological problems, enzyme inhalation, kidney disfunction and diminished learning ability in children (Akhionbare, 2009). Lead concentrations in this study were below WHO standards (0.01ppm) for drinking water.

3.7 Relationships

Correlation matrix from this study (Table 8) showed some significant correlation at P<0.01. pH and Alkalinity (–1.000**) exhibited negative relationship, whereas conductivity and TDS exhibited significant positive correlation, which could be attributed to increase in anthropogenic influence which causes increase in TDS. BOD exhibited positive correlation with Alkalinity (0.960), Sulphate (0.891), Nitrate (0.938), and negative correlations with Salinity (–0.997), Chloride (0.828), Conductivity (–0.985), and TDS (0.986). pH and Alkalinity exhibited significant negative correlation at $p<0.01$, pH also showed negative relationships with Sulphate & Nitrate and positive correlations with the rest parameters.

Significant positive correlation at $p<0.05$ was observed between Cd & Coliform count (table 9). Also significant negative relationship at $p<0.05$ was recorded for Ni & Sulphate. Fe exhibited negative correlation with Chloride (–0.440), Total Alkalinity (–0.963), Conductivity (–0.770), TDS (–0.773), PH (–0.704), and Salinity (–0.826). Ni recorded positive correlations with Chloride (0.980), Conductivity (0.975), TDS (0.974), pH (0.992), Salinity (0.950). The highly significant positive correlation between the concentrations of Cd and Coliform suggests that the association of the two originates from a common source (Untreated waste water and sewage), this is evident within the study area as most dwellings do not have sanitary facilities and as such defecation into the river is the only option.

Several significant relationships at $p<0.01$ and $p<0.05$ were observed amongst heavy metal species in the River water and Fish samples (Table 10). Significant Positive correlation between Ni & Cd (0.867*). Also, significant negative correlations were recorded between Fe &Pb (–0.946**), Cd & Cr (–0.870*), Cd & Zn (–0.869*). The positive correlation between the heavy metals indicates that they are strongly associated. The negative correlation amongst the concentrations may be a result of precipitation of the elements as hydrous metal oxides (El-Sayed & El-Sayed 1980). Pb can bioaccumulate in tissue of fish and human and cause severe health problems such as Brain & Kidney damage, diminishing learning ability in children, also Cd can bioaccumulate in tissues of fish and human and cause several health impact such as weakened immune system and infertility. (Akhionbare, 2009)

3.8 Water Quality Index

The WQI values for three (3) stations in the watershed ranges from 768 to 1064 as shown in Tables 8-10 and therefore, can be categorized as types V (Table 1) "Unsuitable water". The high value of WQI at this station has been found to be mainly due to the higher values of TDS, BOD, Chloride, Nitrate and Sulphate, where it was found that there is a very high correlation coefficient between them.

Table 8. Relationship between variables in water across the three sampling points

	T. alkalinity	Chloride	T. hardness	Sulphate	Nitrate	Conductivity	TDS	pH	Salinity	BOD	C. count
Talkalinity	1										
Chloride	-0.952	1									
Thardness	-0.472	0.18	1								
Sulphate	0.983	-0.992	-0.301	1							
Nitrate	0.802	-0.581	-0.905	0.678	1						
Conductvty	-0.994	0.912	0.568	-0.956	-0.864	1					
TDS	0.993	0.910	0.572	-0.954	-0.867	1.000**	1				
pH	-1.000**	0.948	0.485	-0.980	-0.811	0.995	0.995	1			
Salinity	-0.979	0.870	0.642	-0.925	-0.907	0.996	0.996	0.982	1		
BOD	0.960	-0.828	-0.701	0.891	0.938	-0.985	-0.986	-0.964	-0.997	1	
Ccount	-0.470	0.718	-0.556	-0.625	0.15	0.368	0.364	0.458	0.281	-0.203	1

**. Correlation is significant at the 0.01 level (2-tailed). All variables in mg/l except Conductivity (µScm), Salinity (psu) and Coliform count (cfu/m)

Table 9. Correlation (r) matrix between heavy metals and water variables across the 3 sampling points

	TAlkalinity	Chloride	THardness	Sulphate	Nitrate	Conductivity	TDS	pH	Salinity	BOD	Ccount
Fe	0.693	-0.44	-0.963	0.584	0.986	-0.770	-0.773	-0.704	-0.826	0.868	0.310
Ni	-0.994	0.98	0.370	-0.997*	-0.730	0.975	0.974	0.992	0.950	-0.922	0.567
Cd	-0.444	0.697	-0.580	-0.602	0.179	0.341	0.336	0.431	0.253	-0.174	1.000*

*Correlation is significant at the 0.05 level (2-tailed)

Plate 1. Activities at sampling station 1

Plate 2. Activities at sampling station 2

Plate 3. Activities at sampling station 3

Table 10. Correlation (r) matrix between heavy metals in water with heavy metals in fish species

	Fe	Pb	Ni	Zn	Cd	Cr
Fe	1					
Pb	-0.946**	1				
Ni	-0.522	0.319	1			
Zn	0.768	-0.639	-0.643	1		
Cd	-0.609	0.513	0.867*	-0.869*	1	
Cr	0.716	-0.726	-0.800	0.670	-0.870*	1

**Correlation is significant at the 0.01 level (2-tailed)
*Correlation is significant at the 0.05 level (2-tailed)

4. SUMMARY

The levels of Chloride, Sulphate, Nitrate, TDS and BOD were very high in the River Okpoka. The levels of heavy metals, especially Ni, Cd and Fe and microbiological parameter (Total coliform count) measured were high in water of the River Okpoka. Industrial and domestic activities were some of the factors that contributed the elevated levels of these variables in the water of the current study, the availability of these heavy metals were in these order; Ni>Cd>Fe>Pb, Zn, Cr, & Hg . The levels of heavy metals in sediment of River Okpoka were significantly higher than those in the water sample, especially those of Fe, Zn and Cr. Heavy metals concentrations also varied across sampling station with Fe and Cr recording highest values at sampling station two (2) which is downstream and harbouring several activities such as sand mining and loading, indiscriminate solid and human waste disposal by riparian communities and fishing.. The order of availability of the heavy metals species in the sediment were as follows; Fe>Zn>Cr>Pb>Ni, Cd, & Hg. The levels of heavy metals in different fish samples (Tilapia sparrmanii, Mugilcephalus and Clariasgariepinus) of River Okpoka were significantly higher than those in the water sample, especially those of Cr, Fe and Ni. Heavy metals concentrations also varied amongst fish species with Tilapia sparrmanii recording highest values. The order of availability of the heavy metals species in the fishes were as follows; Cr>Fe>Ni>Zn>Cd>Pb>Hg, while the order of bioaccumulation in fishes were; Tilapia>Catfish>mullet. This study also recorded significant positive and negative correlation between water quality variables, microbiological parameters and heavy metals. The high counts recorded for the microbiological parameters depict abundance and considerable level of decomposition and biodegradation activities in the River. The water quality index (WQI) indicates the deterioration of water quality, water quality is almost always threatened or impaired; conditions usually depart from natural or desirable levels. Consumption of this water by riparian communities could pose severe health consequences.

5. CONCLUSIONS

It can be concluded that point sources of pollution are responsible for the pollution of Okpoka River during the dry season during which this study was carried out. While rainfall may dilute and weaken the effects of point source pollution, it also increases the contribution of non-point source or diffuse pollution through land runoff from riparian communities & agricultural fields and leachates from refuse dumps. The result of the analyses for most parameters did not show the expected trends in water quality from upstream to downstream sites on the river along the 3 sampling sites. It is expected that the concentration of most parameters downstream should be higher than midstream or upstream. This is not the case in this study. This clearly confirms that the main source of contamination Okpoka River is point source pollution. Nevertheless, nonpoint source such as diffuse pollution from agriculture, urbanization, construction etc. add to contamination of the river. River Okpoka is polluted along the river course and poses severe health risk to several riparian communities who rely on the river primarily as their source of domestic water.

ACKNOWLEDGMENTS

We wish to express our gratitude to the Petroleum Training Institute, Warri and the Federal University of Technology Owerri for access to their facilities that made this study a success.

REFERENCES

1. Ajao, E.A. and Fagade, S. O (2002). The ecology of Neritinaglabratain Lagos Lagoon, Nigeria. Arch. Hydrobiol., 119(3): 339-350.

2. Akhionbare, S. M. O (2011). Factors in the migration of Heavy metals in the Otamiri River system.International Journal of science and nature.2(4): 856–860.

3. Akhionbare, S M O; (2007). Heavy metal distribution in natural water sources in the owan area of Edostate Nigeria inter Res J. in Engr. Sc and Tech. IRSSEST. 4(2): 88-95.

4. Akhionbare, S.M.O. and Akhionbare, W.N. 2005. The Composition, Seasonal Variation and Distribution of Bacteria in a Small West African Lake.InternationalResearch Journal in Engineering, Science and Technology (IREJEST). 2 (1): p 102-110.

5. Ansa, E.J., (2005). Studies of the benthic macrofaunaofthe Andoni flats in the Niger Delta Area of Nigeria. Ph.D Thesis University of Port Harcourt, PortHarcourt, Nigeria, pp: 242.

6. APHA (2012). Standard methods for examination of water and waste water.(22nd, Edn.) APHA, AWWA, WPCF, Washington DC, USA.

7. Boyd, C.E. and Lichtkoppler, (1979). Water quality management in fishpond, Research and Dev. Series No.22. International centre for Aquaculture and Agriculture (I.C.A.A.), Experimental station Auburn University, Alabama, pp: 45-47.

8. Braide, S.A., Izonfuo, W. A. L., Adiukwu, P. U., Chindah, A. C. and Obunwo, C. C (2004). Water Quality of Miniweja stream, A Swamp forest stream receiving non-point source waste discharges in Eastern Niger Delta, Nigeria.Scientia Africana, 3(1): 1-8.

9. Chamie J. (2004). Statement to the commission on population and development (Thirtyseventh session). Population Division, Department of Economic and Social Affairs.United Nations.

10. Charbonneau, C.S and Nash, T (1993). Contaminants program, Mingo National Wildlife Refuge (Region 3), contaminants survey results. U.S. Fish and Wildlife Service, 608.

11. CCME. (2001). Canadian water quality guidelines for the protection of aquatic life: Canadian Water Quality Index 1.0 Technical Report. In Canadian environmental quality guidelines. 1999. Winnipeg.

12. Defew, L. H., Mair, J. M and Guzman, H. M (2005). An assessment of metal contamination in mangrove sediments and leaves from punta mala bay. Poll Bulletin (50): 547–552.

13. Dangana, L.B., (1985). Hydrogeomorphological controls of the mangrove environment In: Proceedings of a Workshop on the mangrove ecosystem of the Niger Delta. University of Port Harcourt. pp: 357.

14. Dublin-Green, C.O., (1990). Seasonal variations in somephysico-chemical parameters of the Bonny Estuary, Niger Delta. NIOMR Technical Paper, 59: 21-25.

15. Eisler, R. (1993). Zinc hazards to fish, wildlife and invertebrates: a synoptic review. U.S. Fish Wildl. Serv. Biol. Rep., 85 (1.10): 90.

16. Ekeh, I.B. and Sikoki, F.D. (2003). The state and seasonalvariability of some physico-chemical parameters inthe New Calabar River, Nigeria.Supp.adActaHydrobiol., 5: 45-60.

17. Greany, K. M. (2005). An assessment of heavy metal contamination in the marine sediments of Las Perlas Archipelago, Gulf of Panama, M.sc. thesis, School of Life Sciences Heriot-WattUniversity, Edinburgh, Scotland, 114pp.

18. Hassan, R., Schole, R. and Ash, N. (eds.) (2005), Ecosystems and human well-being: current state and trends. Vol. 1 Island press, Washington DC 917 pphttp://www.millenniumassessment.org/en/Condition.aspx#download Accessed: 25 August, 2010.

19. Horton, R. K. (1965). An Index Number System ForRating Water Quality.Journal-Water Pollution Control Federation. 37:300-305.

20. Jewitt, G. (2002). Can Integrated Water Resources Management sustains the provision of ecosystems goods and services? Physics and Chemistry of the Earth 27: 887-895.

21. Moses, B.S., 1983. Introduction to Tropical Fisheries.Ibadan University Press, UNESCO/ICSU Part, pp: 102-105.

22. Payne, A.I., (1976). The relative abundance and feeding habits of the grey mullet species occurring in an estuary in Sierra Leone, West Africa. Marine Biol., 35: 275- 286.

23. Ramakrishnalah, C.R., Sadas-hivalah, C. and Ranganna G.(2009). Assessment of Water Quality Index for the Groundwater in TumkurTaluk, Karnataka state, India, E Journal of chemistry, 6(2):523-530.

24. Rompala, J.M., Rutosky, F.W., Putnam, D.J. (1984). Concentrations of environmental contaminants from selected waters in Pennsylvania. U.S. Fish Wildl. Serv. Rep., State College, Pennsylvania., pp: 102.

25. Sorenson, E.M. (1991). Metal Poisoning in Fish. CRC Press, Inc. Boca Raton, Florida, pp: 119-174.

26. Swingle, H.S., (1969). Methods of analysis for waters organic matter and pond bottom soils used in fisheries research. Auburn University, Auburn, Alabama, pp: 119.

27. Tam, N. F. Y., Wong, W. S (2000). Spatial variation of heavy metals in surface sediments of Hong Kong Mangrove swamp. Environmental Pollution 110:195–205.

28. World Health Organization (WHO) (2008): International standards for drinking water World Health Organization. Geneva, Switzerland. 2nd Ed.

Chapter 6

Hydrological Modeling of Aquifers and Their Ground Water Potentials: Implications for Water Resources Planning and Management in Parts of Obio/Akpor L.G.A, Rivers State, Nigeria

Vincent Ezikornwor Weli[1], Jimmy Adegoke[2], Ogah Celestine Ndidi[1]

[1]Department of Geography and Environmental Management, University of Port Harcourt, Port Harcourt, Nigeria

[2]Department of Geosciences, University of Missouri, Kansas City, USA

ABSTRACT

This study examined the hydrological modeling of aquifers and their ground water potentials for the purposes of water resources planning and management. This was done using the electrical resistivity method employing the schlumberger electrode configuration at randomly selected stations to obtain the thicknesses and resistivities of each layer and depth to the presumably conglomeratic sand stone and its resistivity. Findings showed that the top soil layer resistivity values vary from 59.3 to 248.4 ohm-m and thickness of 0.6 to 3.9 m. The second layer has resistivity values ranging from 45.0 to 743.5 ohm-m and a thickness range of 1.5 to 13.8 m. The wet sand is characterized by resistivity values ranging from 144.8 to

1930.2 ohm-m and a thickness range of 3.8 to 65.8 m. The conglomeratic sand/sand stone has resistivity values ranging from 55.8 to 7719.8 ohm-m. The depth to this bottom layer varies from 6.6 to 89.5 m. Findings indicate that the entire profile is a sedimentary formation represented by lithological units of sand and clayey sand which make for a good groundwater potentials. However, the groundwater potential zones of the study area in terms of transmisivity revealed four distinct classes representing "very good" (Mgbuosimini, Rumuigbo, Okporo, Rumuomasi and Rumuodara), "good" (Alakahia, Rumuodomaya, Oginigba and Rumuola), "moderate" (Aluu, Rumuekeni, Rumuokoro, Rumuobiakani and Rumueme), and "low" (Ogbogoro, Ozuoba, Akpajio, Elelenwo, Eliozu, Rumuepirikon, Rumuokwuta, Rumuebekwe and Rumurolu) groundwater potential in the area. Well logging should therefore be incorporated in borehole development process for safe and sustainable yield of groundwater in Obio/Akpor.

KEYWORDS

Aquifers, Water Resources, Sedimentary Formation, Underground Water and Obio/Akpor

1. INTRODUCTION

The increased interest in recent years in underground sources of water has led to a need for more intensive studies of the geometry and properties of aquifers [1] . This is because over half of the world's population depends on groundwater for drinking water supplies. For instance, in the United Kingdom about 30% of the public water supplies are derived from groundwater, in the United states of America about 50%, Denmark 99% [2]; Germany 70% [3] ; and in Tanzania, 30% [4] depend on underground water resources. Geophysics has played a useful part in such investigations for many years and improvements in instruments and the development of better methods is resulting in a widening of its applications. A geophysical model created can be used to support other studies which involve, primarily one-, two-, and three-dimensional modelling. Resistivity imaging is one of the geophysical surveys which have been used to map groundwater contamination and it is widely used for environmental surveys [5] - [8] . This technique has been used to determine the subsurface resistivity anomalies and recently it has become popular for the investigation of water movement in vadose zone. The growing population in Obio/Akpor has been projected to be around 2 million by 2018. And as a result, water supply must be efficient enough to meet the water demand of the inhabitants in the area.

The ground-water hydrology of some parts of Obio/Akpor local government area, however, is not properly understood. Most previous studies on groundwater potential in the area have tended to concentrate on groundwater quality with reference to contamination [9] [10] . There is a variable degree of knowledge and management practice, ranging from almost no data and no action to sound models about aquifer behavior [11] . In addition, little consideration has been given to evaluate the potential for new groundwater uses to affect existing groundwater users. This lack of understanding could hinder efficient and prudent ground-water development. A thorough study to characterize and quantify the ground-water flow system in the area is needed to address these gaps in the study area. This study therefore is an attempt to hydrologically model the aquifers to determine groundwater flow direction and identification of major hydro geologic units.

2. GEOGRAPHY AND GEOLOGY OF OBIO/AKPOR

The study area Obio Akpor L.G.A, Rivers State is delineated between Latitude 4°30' and 4°55'N and longitudes 6°55' and 7°00'E (Figure 1) within the Niger Delta region of Nigeria. The Niger Delta area has a tropical rainforest climate, with a distinct wet (April-October) and dry (November-March) seasons. The mean annual rainfall ranges from 2000 mm (inland) to over 4000 mm at the coast and the area is characterized by five major geomorphic units namely dry flatland & plains, Sombreiro-Warri deltaic plains with abundant freshwater back swamps, freshwater swamps, meander belts & alluvial swamps, saltwater or mangrove swamp and active/ abandoned coastal ridges [12] . The proximity of the aquifers to the surface, flat topography, high annual rainfall, and permeable soil media contributes to insignificant runoffs in the site, and implies that the total precipitation goes into storage. This enhances decomposition activities by bacteria and fungi and leaching of contaminants into the aquifer. The area is geologically composed of various Quaternary deposits that overlie the three main stratigraphic units of Benin, Agbada and Akata. The Benin Formation (Oligocene to Recent) is the aquiferous formation in the study area and is exploited for groundwater supplies [12] . Although a depth of 100 m is most exploited, about 300 m depth has been exploited for water [13] . The Benin Formation consists essentially of massive and highly porous sands and gravels with a few thin clay intercalations. The uppermost section of the Benin Formation is the Quaternary deposits of about 40 - 150 m thick and comprises of rapidly alternating sequences of sand and silt/clay with the later becoming increasingly more prominent seawards [14] .

The quality of the groundwater is generally good. The occurrence of iron rich water in the Niger Delta is a well-known problem. Although the iron problem is more prevalent in the freshwater and saltwater mangrove swamps, it can sometimes constitute a major problem in some isolated localities in the dry deltaic plains. Unfortunately, the major factors controlling the lateral and vertical occurrence and distribution of high iron content groundwater in the Niger Delta have not been confirmed [15] . Obio/Akpor features a tropical monsoon climate with lengthy and heavy rainy seasons and very short dry seasons [16]. Only the months of December and January truly qualifies as dry season months in the area, although rainfall is experience in all the months. The harmattan, which climatically influences many cities in West Africa, is less pronounced in the area. The region's heaviest precipitation occurs during September with an average of 367 mm of rain. December on average is the driest month of the year with an average rainfall of 20 mm. Temperatures throughout the year in the city are relatively constant, showing little variation throughout the course of the year. Average temperatures are typically between 25°C - 28°C in the city.

3. METHODOLOGY

The Instruments used for data acquisition include Resistivity meter, 12 Volts DC battery, Four steel electrodes, Four realms of insulated conductor cables, Two measuring tapes, Two 250 g hammer, Four clips, Compass and GPS, Calculator Data sheet, Pencil, Pen, Eraser and sharpener. The electrical resistivity survey was done by dividing the study area into four major groups with each group given its own strike using the Schlumberger electrode configuration at four selected stations for striking within the study area to obtain the layer parameters (resistivities and thicknesses) upon which generations of maps were made. The Schlumberger was used to ensure deep penetration and for logistics of limited man power in the field. The following constitutes the four major groups: Group A (NE-SW

Strike) Choba, Ozuoba, Rumuekini, Rumuekini, Mgbuoba, Aluu, Alakahia, Rumuosi. Group B (NE-SW Strike) Rukpokwu, Rumuchiolu, Rumugbo, Rumuodomaya, Atali, Rumuokoro, Elelenwo, Eliozu, Apajo. Group C (NW-SE Strike) Rumuokwuta, Rumueme, Rumuepirikom, Mgbuosumini, Rumuola, Rukpakolusi. Group D (NW-SE Strike), Rumuomasi, Oginigba, Rumuorolu, Rumuobiakani, Rumuogba, Rumuiebekwe, Okporo, Apajo. The 30 VES locations studied were made into 4 profiles (AB, CD, EF and GH respectively) with definite orientations, transversing different geological formations to project a 2-D subsurface geo-electric section along that profile in order to indicate the distinct subsurface layers in the study area. The layers are the top soil, clayey sand, sandy clay, wet sand and presumably conglomeratic sand or sand stone. The geo-electrical resistivity survey of the area was carried out varying the half-current electrode spacing from 1 to 60 m with a total of 30 VES stations (Figure 1). The electrical resistance measurements were made with the rectified resistivity meter. The apparent resistivity values were computed using Equation (1).

$$\rho = \frac{\Delta \upsilon}{I} * \pi/2p * \left(r^2 - p^2\right) \tag{1}$$

where AB/2 = r and MN/2 = p and $\frac{\pi}{2p} * \left(r^2 - p^2\right)$ is the geometric factor, G.

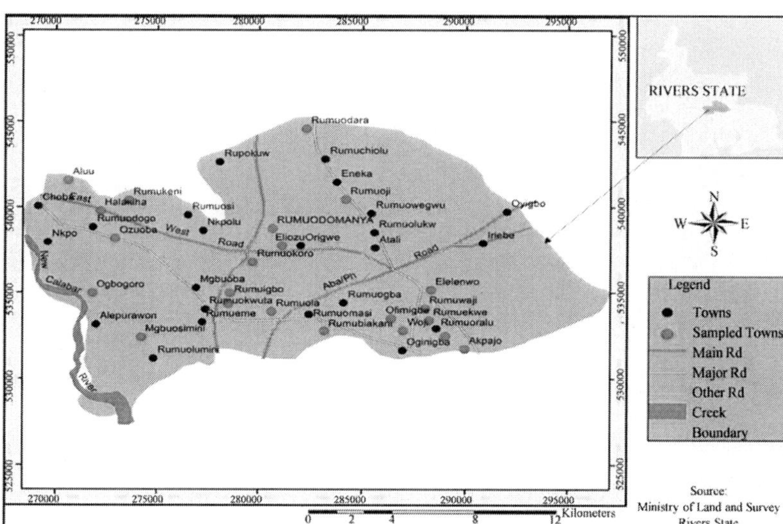

Figure 1. Obio/Akpor, showing sampled towns.

The apparent resistivity curves were plotted with half current electrode spacing (AB/2) along the abscissa and apparent resistivity values along ordinate on a bi-logarithmic graph sheet. The apparent resistivity curves were quantitatively interpreted by partial curve matching using 2-layer master curve and a set of auxiliary curves. The resultant starting models were subjected to 1-D inversion using WINREST software. The flow net of the study area was done in order to determine the direction of flow of the groundwater with respect to topography. The values were contoured using SURFER 8 software with the easting, northing, and elevations represented in the X, Y and Z components respectively. The layering parameters in terms of layer Resistivities and thicknesses from the interpretation of the apparent resistivity curves were used to construct geo-electric sections along profile lines. The geo-electric section attempts to correlate different layers at each VES point. These Geo-electric profiles were generated using Corel draw software package. The geo-electric profiles are the geo-electric section of each VES location united into one profile with a definite directional orientation. The ordinate "X" axis represent resistivity and the abscissa "Y" axis represent the depth in metres below ground level. The isopach, overburden coefficient of anisotropy and iso-resistivity maps at each VES point was done using SURFER 8 software. The easting, northing and the top aquifer layer resistivity represent the X, Y and Z components respectively. This is because the lower the resistivity of the top aquifer the higher its groundwater potential. The overburden coefficient of anisotropy (Λ) was calculated for each VES station using the layer Resistivities and thicknesses as shown in Equation (2).

$$\Lambda = \sqrt{\ell t/\ell_L} \qquad (2)$$

where;

Longitudinal unit conductance, $S_i = h_i/\ell_i$.

Transverse unit resistance, $T_i = h_i \cdot \ell_i$.

Longitudinal resistivity, $\ell_L = h_i/s_i$.

Transverse resistivity $\ell_t = T_i/h_i$.

The empirical relation between aquifer parameter and resistivity as used by [17] and used in this study are established for transforming resistivity distribution into permeability and transmisivity of the aquifer according to the Equation (3) below;

$$K = 0.0538e^{0.0072\rho} \quad (3)$$

where;

K = Hydraulic conductivity of Aquifer and

ρ =Apparent Resistivity of Aquifer.

4. RESULT AND ANALYSIS OF FINDINGS

Findings indicate that the flow net of the study area has a series of highs and lows. The areas with the lows (0 - 2.4 m) can be found around NW of Akpajo, W of Akpajo and N of Rumuokoro and represent low elevation points. Areas with intermediate values (2.5 - 3.1 m) can be found around S of Rumuokoro, NE of Okporo, E of Ozuoba, N of Ozuoba and SE of Ozuoba. While areas with high values (≥3.2 m) can be seen around NW of Aluu, S of Aluu, SW of Ozuoba, NW of Ozuoba, S of Eliozu and Okporo and SE of Rumuokoro and represent high elevation points. The highs and the lows are indicators of the elevation and water potential at the water table [18] and the slope of the water table generally follows the contours of the surface topography [19] . Therefore the groundwater will be expected to move from the highs through the intermediary to the lows in the study area (Figure 1).

There are different curve types inferred during the geophysical interpretation of the geo-electric parameters. The sounding curves are typified by AK, A, K, KH, AH, Q, KK, H and AA curve types (Table 1).

The curve types of the study area are expected to follow the general trend as described below:

K type—The "K" curve is characterized by decreasing resistivity values with depth. The resistivity value also decreased in the deeper aquifer. This is a typical representation in Ozuoba. H type—The "H" curve type is a typical representation in Mgbuosimini where the resistivity value of the first layer is greater than the resistivity of the second layer and resistivity of the second layer is less than that of the third layer. Q type—The "Q" curve shows a continuous decrease in resistivity values from the first layer to the under laying layers. This is a typical curve representation in Rumuigbo. A type—The "A" curve type shows a continuous increase of resistivity values from the first layer down to the under laying layers. This is a typical representation of Aluu, Ogbogoro, Rumuopirikon, Rumuokwuta, Rumuola, Rumuobiakani and Rumuomasi. Others are AK (represented in Alakahia, Rumuodara, Rumuoomaya and Oginigba), KH (represented in Rumuekeni, Akpajio, Elelenwo, Rumuibekwe, Rumurolu and Rumueme), AH (represented in Eliozu), KK (represented in Rumuokoro) and AA (represented in Okporo) curve types. The units of y and x coordinates of the Figures 2-4 below are in meters.

The figures below show typical sounding curves for different stations. However because of space, very few were shown here. Values for x are measured in meters and y coordinates are measured in ohm-meters as shown (Figures 5-7).

The isopach map of the overburden revealed overburden thicknesses between 0 and 46 m. The overburden as used in this context refers to all materials above the presumably conglomeratic sand/ sand stone. Findings showed that the highs around E and S of Eliozu have thick overburden thicknesses (≥28 m) while the lows as seen around SW of Rumuokoro, N and NW of Rumuokoro, N and NE of Okporo, E and SE of Okporo have relatively thin overburden (0 - 9 m). Areas with intermediate overburden thicknesses (10 m - 27 m) include E and SE of Rumuokoro, N and SW of Eliozu and W and NW of Okporo. Consequently areas with highs may have a minimal groundwater potential due to a high drill depth for borehole while good and better groundwater yield is expected in areas with average and thick overburden thicknesses respectively due to a fair and smaller drill depth for borehole development in the study area.

The aquifer resistivity of the VES points indicates that the lows (≤300 ohm-m) are seen around E of Aluu, NE of Akpajo and N of Eliozu and as such are expected to be good aquifers in contrast to the highly resistive areas (≥700 ohm-m) which include S of Ozuoba, NE of Okporo and SE of Okporo. Areas with intermediate resistivity value ranging between 300 and 600 ohm-m around N of Okporo, S of Okporo and NE of Rumuokoro are expected to have fair aquifer qualities (fair groundwater yield).

Table 1. Curve types from interpretation of VES data where ρ represent resistivity.

Area	Curve type	Curve characteristics	No of geoelectric layers
Alakahia, Rumuodara, Rumuodomaya, Oginigba	AK	$\rho_1 < \rho_2 < \rho_3 > \rho_4$	4
Aluu, Ogbogoro, Rumuepiriekom, Rumuokwuta, Rumuola, Rumuobiakani, Rumuomasi,	A	$\rho_1 < \rho_2 < \rho_3$	3
Ozuoba	K	$\rho_1 < \rho_2 > \rho_3$	3
Rumuekeni, Akpajo, Elelenwon, Rumuebekwe, Rumurolu, Rumueme	KH	$\rho_1 < \rho_2 > \rho_3 < \rho_4$	4
Eliozu	AH	$\rho_1 < \rho_2 < \rho_3 < \rho_4$	4
Rumuigbo	Q	$\rho_1 > \rho_2 > \rho_3$	3
Rumuokoro	KK	$\rho_1 < \rho_2 > \rho_3 > \rho_4$	4
Mgbuosimini	H	$\rho_1 > \rho_2 < \rho_3$	3
Okporo	AA	$\rho_1 < \rho_2 < \rho_3 < \rho_4 < \rho_5$	5

The overburden coefficient of anisotropy showed that the lows (≤1.15) have relatively lesser values compared to the highs (≥1.4). The areas with high anisotropy values can be located around SW of Rumuokoro while areas with low anisotropy are located around East of Aluu, NE of Eliozu, SE of Rumuokoro, SE of Okporo and NW of Akpajo. The areas characterized with intermediary values of overburden coefficient of anisotropy can be found around S of Ozuoba, S of Rumuokoro and S of Okporo (1.2 - 1.35 m). According to [20] , a higher coefficient of anisotropy implies higher permeability. Thus areas such as SW of Rumuokoro have good groundwater potential in contrast to East of Aluu, NE of Eliozu, SE of Rumuokoro, SE of Okporo and NW where anisotropy is relatively low.

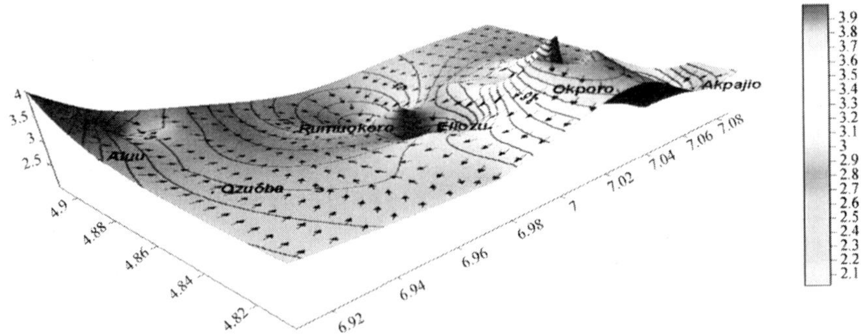

Figure 2. Vector map representing groundwater flow of the area.

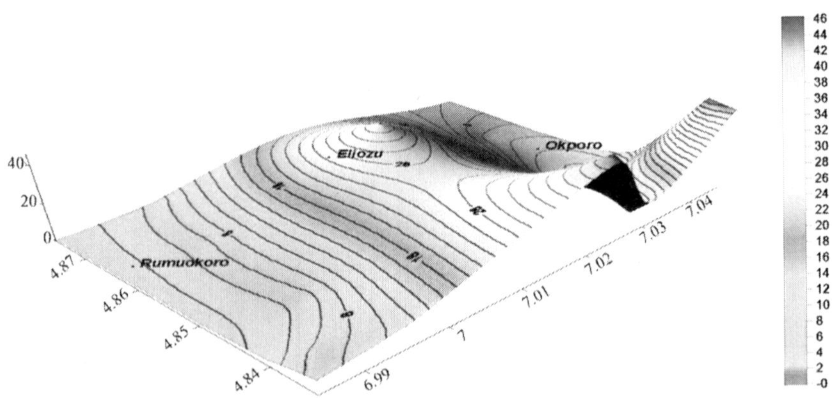

Figure 3. 3D map of overburden.

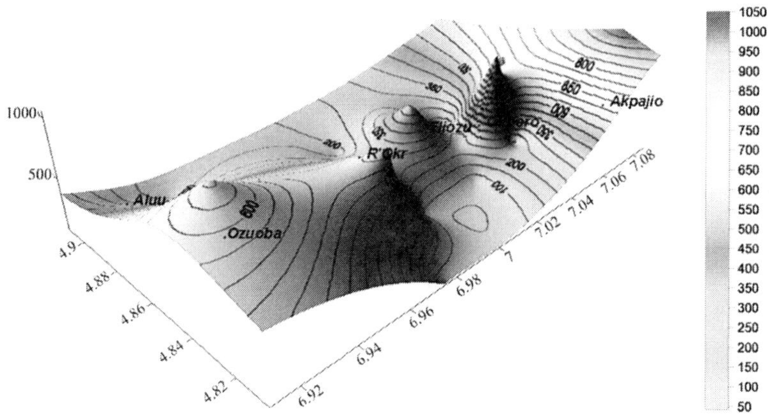

Figure 4. Isoresistivity map of the study area.

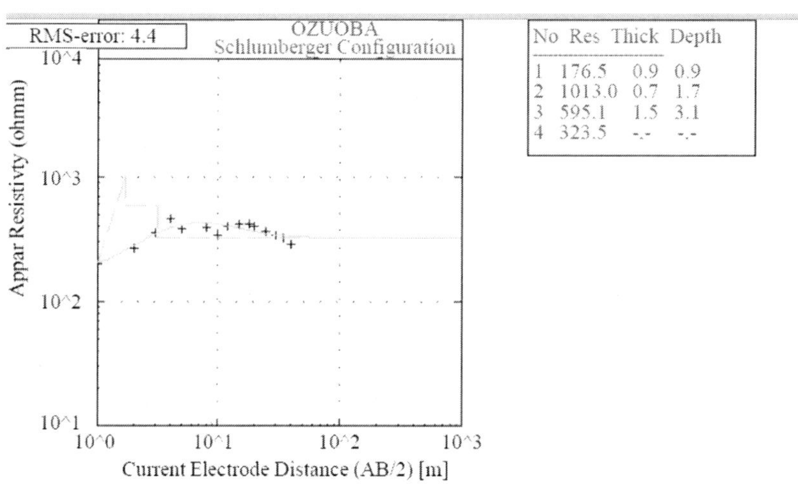

Figure 5. Typical sounding curves for Uzuoba.

Findings revealed that the Isopach map of Top Aquifer is characterized by a systematic increase towards the South (SW precisely, 15 m to 45 m) and a decrease towards areas around the North (precisely NE area, below 15 m). This implies that areas around the SW with large Top Aquifer thickness will be best suited for prolific borehole yield due to their high ground water recharge. While areas around the NE will have poor ground water recharge.

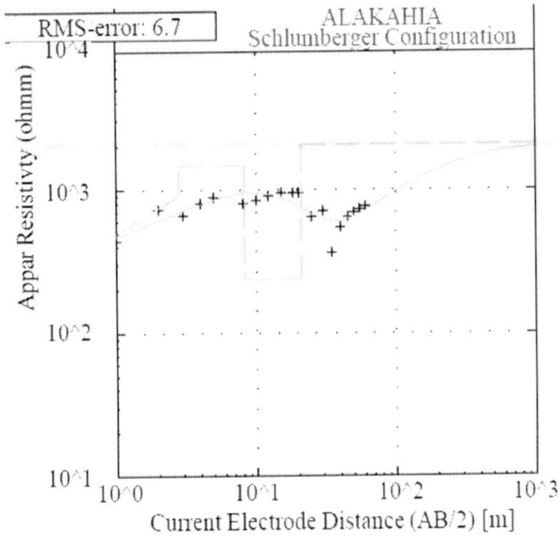

No	Res	Thick	Depth
1	415.3	0.7	0.7
2	900.7	2.1	2.8
3	1469.8	5.5	8.3
4	236.8	12.8	21.2
5	2046.7	-.-	-.-

Figure 6. Typical sounding curves for Alakahia.

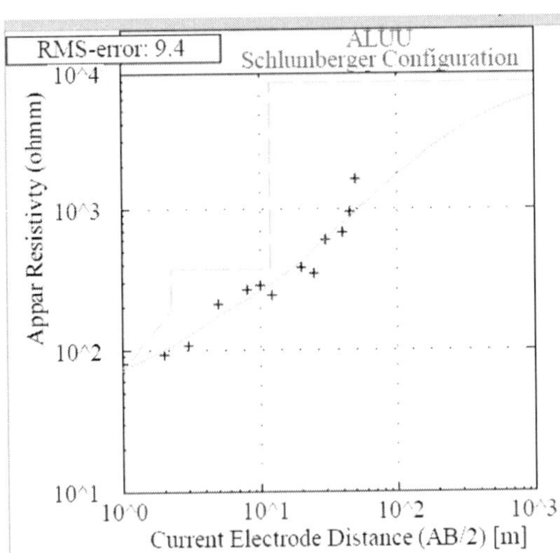

No	Res	Thick	Depth
1	69.0	1.0	1.0
2	193.3	1.3	2.3
3	370.2	9.6	11.8
4	7902.9	-.-	-.-

Figure 7. Typical sounding curves for Aluu.

The profile AB comprises of 5 VES points located at Ozuoba, Alakahia, Rumuekeni, Aluu and Ogbogoro in NE-SW direction. The top soil layer resistivity values vary from 69 - 415 ohm-m and thickness of between 0.6

and 1.0 m. The second layer has resistivity values ranging from 193.3 to 1013.0 ohm-m with a thickness range of 0.3 to 2.7 m. The wet sand is characterized by resistivity values ranging from 179.3 to 1469.8 ohm-m and thickness range of 1.5 to 9.6 m. The conglomeratic sand has resistivity value range of 169.3 to 7902.9 ohm-m. The depth to this bottom layer varies from 3.1 to 21.2 m.

The profile CD comprise of 7 VES points located at Eliozu, Elelenwo, Akpajo, Rumuokoro, Rumuigbo, Rumuodomaya and Rumuodara in the E-W direction. The top soil layer has resistivity value range from 56.3 to 2373.1 ohm-m and thickness range of 0.5 to 3.6 m. The second layer has resistivity range of 172.4 to 5528.8 ohm-m and a thickness range of 1.0 m to 32.6 m. The third layer also has a resistivity range of between 210.4 and 7410.o ohm-m and a thickness range of 0.7 and 13.8 m. The wet sand is characterized by a resistivity value range of 79.7 to 10554.1 ohm-m and a thickness range of 1.0 m to 13.3 m.

The conglomeratic sand/sand stone has a resistivity value range of 367.3 to 5770 ohm-m. The depth to this bottom layer varies from 6.7 to 36.5 m.

From the profile findings it can be inferred that the low resistivity anomalies may be due to the presence of clay while the higher resistivity maybe due to the presence of sand. Based on the geo-electrical investigations of the study area, the entire profile is a sedimentary formation represented by lithological units of sand and clayey sand which are good groundwater potentials. The occurrence of alternating sequences of clay and sand deposits call for proper and careful well siting, drilling and development by use of adequate borehole technology despite great potential for groundwater resource in the study area.

Findings indicate that areas where we have high resistivity values turns out to be areas where we have low transmisivity values and vice versa. Some of the areas have high depth to aquifer (VES 21, VES 8, VES 22, and VES 20) which may warrant a high drilling depth for borehole development in the areas. Whereas some of the areas have shallow aquifer (VES18, VES7, VES4, VES5, VES3) which may be easily polluted by contaminated surface runoff in the area. Moreso, some of the areas have large aquifer thickness (VES22, VES17, VES9, VES1, VES16, VES13) which may warrant a high groundwater content. Areas with small aquifer thickness as representedbyVES3, VES4, VES21, VES23, VES19, and VES5 may be expected to have low groundwater content. Also from the contour and 3D transmisivity maps of the probable aquifer distribution represented by Figure 8 and Figure 3 respectively, it is observed that areas around the SW and due South of the study area have high Transmisivity values (≥4.5 m^2/s). These areas are therefore expected to have good groundwater potentials in form of high groundwater yield due to their presumably high discharge. However the groundwater potential zones of the study area in terms of Transmisivity reveals four distinct classes representing "very good" (Mgbuosimini, Rumuigbo, Okporo, Rumuomasi and Rumuodara), "good" (Alakahia, Rumuodomaya, Oginigba and Rumuola), "moderate" (Aluu, Rumuekeni, Rumuokoro, Rumuobiakani and Rumueme), and "low" (Ogbogoro, Ozuoba, Akpajio, Elelenwo, Eliozu, Rumuepirikon, Rumuokwuta, Rumuebekwe and Rumurolu) groundwater potential in the area.

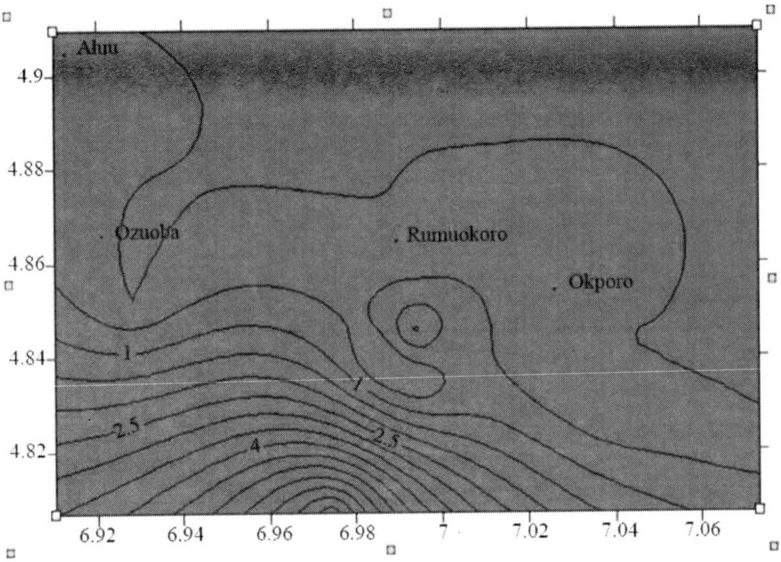

Figure 8. Contour map of top aquifertransmisivity.

5. CONCLUSION

The low and higher resistivity anomalies found were attributed to the presence of clay and sand respectively. Similarly, the entire profile is a sedimentary formation represented by lithological units of sand and clayey sand which are good groundwater potentials. The occurrence of alternate sequence of clay and sand deposits demands for proper and careful well sitting, drilling and development despite great potentials for groundwater resource in the study area. Good groundwater yield can be located around SW of the study area, SW of Rumuokoro, N of Rumuokoro, E of Aluu, NE of Akpajo and N of Eliozu. This is because they are characterized by high anisotropy coefficient (i.e. high permeability), high aquifer thickness, high transmisivity and low resistivity. This implies that the aquifers in these areas are expected to be saturated with groundwater and also transmit water at a high rate and capable of supporting groundwater exploration for large scale water supply.

6. RECOMMENDATION

1) Well logging should therefore be incorporated in borehole development process for safe and sustainable yield of groundwater in the area.

2) It is therefore suggested that boreholes be drilled at Mgbuosimini, Rumuigbo, Okporo, Rumuomasi and Rumuodara, in order to sustain regular water supply in the study area.

3) Pollution Source Release Control should be monitored in the area.

4) Period Groundwater Monitoring is needed.

5) Aggressive Public enlightenment campaign is highly needed.

6) Studies are needed in the following areas.

    a) Determination of rates and distribution of groundwater yield.

    b) Identification of contaminants of emerging concerns (CECs) in the groundwater resources of the region.

7. HIGHLIGHTS OF THE STUDY

- Good ground water yield was found around Rumuokoro, Aluu, Akpajo and Eliozu.

- The ground water flow direction is towards NW and W of Akpajo and N of Rumuokoro.
- Clay and sand caused low and high resistivity anomalies respectively.
- Well logging in borehole development process is proposed.
- Alternative source of water supply is needed.

ACKNOWLEDGEMENTS

We wish to acknowledge the schlumberger oil servicing company in Port Harcourt for the use of their equipment throughout the period of this study.

REFERENCES

1. Asry, Z., Samsudin, A.R., Yaacob, W.Z. and Yaakub, J. (2012) Groundwater Exploration Using 2-D Geoelectrical Resistivity Imaging Technique at Sungai. Udang, Melaka. Journal of Earth Science and Engineering, 2, 624-630.

2. Tebbuth (1992) Principles of Water Quality Control. Elsevier Science and Technology Books.

3. Trauth, R. and Xanthopoulos, C. (1997) Non Point Pollution of Ground Water in Urban Areas. Water Research, 31, 2711-2718.

4. Masterson, J.P., Carlson, C.S., Massey, A.J. and Walter, D.A. (2009) Hydrogeology and Simulation of Groundwater Flow in the Plymouth-Carver-Kingston-Duxbury Aquifer System South Eastern Massachussets: U.S Geological Survey Scientific Investigation Report.

5. Griffiths, D.H. and Barker, R.D. (1993) Two-Dimensional Resistivity, Imaging and Modelling in Areas of Complex Geology. Applied Geophysics, 29, 211-226.

6. Samsudin, A.R., Hamzah, U., Rafek, A.G. and Yaacup, R. (2000) Application of Geoelectrical Resistivity Imaging for Site Investigation. Proceedings Annual Geological Conference, Bali, 375-380.

7. Samsudin, A.R., Hamzah, U. and Ramli, Z. (2001) Geophysical Investigation of the Olak Lempit Quaternary Basin. In: Geosea Regional Congress, Yogyakarta, 280-282.

8. Hamzah, U., Samsudin, A.R., Yaacup, R. and Ayub, S.M. (2006) Electrical Imaging of Groundwater Aquifer at Banting, Selangor, Malaysia. Environmental Geology, 49, 1156-1162.

9. Nwankwoala, H.O. and Omunguye, M.L. (2012) Geophysical Investigation for Ground Water in Boirikiri and Eastern Bye-Pass Areas of Port Harcourt. The Pacific Journal of Science and Technology, 14, 524-535.

10. Ujile, A.A. (2013) Evaluating Groundwater Contamination Process and Developing Framework for Qualitative Management in Parts of Nigeria.

11. Nwankwoala, H.O. (2011) An Integrated Approach to Sustainable Groundwater Development and Management in Nigeria. Journal of Geology and Mining Research, 3, 123-130.

12. Akpokodje, E.G. (2001) Introduction to Engineering Geology. Pam Unique Publishing Co. Ltd., Port Harcourt, 180-181.

13. Ngah, S.A. (1990) Groundwater Resource Development in the Niger Delta: Problems and Prospect. 6th International IAEC Congress.

14. Etu-Efeotor, J.O. and Akpokodje, E.G. (1990) Aquifer Systems of the Niger Delta. Journal of Mining and Geology, 26, 279-285.

15. Port Harcourt Water Corporation (PHWC) (2014) Urban Water Reform and Port Harcourt Water Supply and Sanitation. African Development Bank Group. Reference: P-NG-E00-007.

16. Weli, V.E. and Worlu, S.O. (2011) Recent Rainstorm Dynamics and Its Implications for Flood Frequency in Sub-Humid Tropical City of Port Harcourt, Nigeria. Journal of Geographic Thought, 12.

17. Singh, K.P. (2005) Non-Linear Estimation of Aquifer Parameters from Surficial Resistivity Measurements: Hydrogeology and Earth System Science Discussion (HESSD). 919-927.

18. Devine, C.E. and McDonell, J.J. (2005) The Future of Applied Tracers in Hydrogeology. Hydrogeology Journal, 13, 255-258.

19. Weis, M. and Gvirtzman, H. (2007) Estimating Groundwater Recharge Using Flow Models of Perched Karsticaquifers. Groundwater, 45, 761-773.

20. Betayanah, A.T. (2006) Use of Electrical Resistivity Methods for Detecting Subsurface Fresh and Saline Water and Delineating Their Interfacial Configurations. A Case Study of the Eastern Dead Sea Coastal Aquifer. Jordan Hydrogeology Journal, 14, 1277-1283.

Chapter 7

Prioritization of Water Resources Management Problems in North Central Nigeria Using Rapid Impact Assessment Matrix (RIAM)

Olayinka Gafar Okeola[1], Khadijat Abdulkareem Abdul Raheem[2]

[1]Department of Water Resources and Environmental Engineering, University of Ilorin, Ilorin, Nigeria

[2]Department of Civil Engineering, University of Ilorin, Ilorin, Nigeria

ABSTRACT

This study employed the Rapid Impact Assessment Matrix (RIAM) to prioritize the water resources management problems in the North Central Nigeria. This was done through the assessment of the status of water resources management in the region, evaluation of existing policy and strategy of water management, identification of the management problems and the prioritization with RIAM. The stakeholders identified water resources management problems, ranked them in other of severity in different categories and also evaluated them using the RIAM techniques in the administered questionnaire. Eleven problems were analyzed based on the physical/chemical, biological/ecological, social/cultural and economic/operational factors using several impact indicators. Scores were assigned, the RIAM models applied and the averages taken to arrive at the final assessment scores. The two major water resources management problems identified are: 1) inadequate funds for further agricultural, hydroelectric, navigation and industrial development; 2) poor data collection and banking. These problems were prioritized by RIAM in order of severity for urgent intervention. The RIAM technique has made a key contribution to the prioritization of water resources management by providing insights into urgent problems according to stakeholders and thus guides the policy maker in appropriate decision making.

KEYWORDS

Water Resources, Rapid Impact Assessment Matrix (RIAM), Stakeholders, Prioritization, North Central Nigeria

1. INTRODUCTION AND LITERATURE REVIEW

Nigeria is assumed to be endowed with abundant water that cannot be exhausted; however, there are increasing indications that contradict this statement. As the nation's population and socio-economic activities increase so also water pollution, water demand for domestic, agriculture, industry and other usage. Moreover, climate change is making Nigeria increasingly vulnerable to frequent droughts and flash floods with attendant loss of life and property. All these are leading to water crises that need urgent measures to reverse. Besides, the importance of water for socio-economic development is globally recognized since accessibility to adequate water supply is central to life and sustainable development. Nigeria like other sub-Saharan countries loses about 5 percent of its GDP annually as a result of under-development and poor management of water resources. This loss value exceeds the total foreign aid and debt relief inflow into the country [1] .

The Federal Ministry of Water Resources (FMWR) is constitutionally empowered to oversee the development and management of Nigeria's surface and groundwater resources including international waters with the following principal functions [2] : formulate appropriate water resources legislation; update and implement the Water Resources Master Plan; formulate and implement national irrigation policy; develop and support irrigated agriculture; coordinate the development and utilization of water resources for irrigation and other purposes; collect, store, analyze and disseminate hydro-meteorological, hydrological and other data among others. The National Water Resources master plan was completed in 1995 with the objective to ensure optimum use of the nation's water resources. It provided for developmental scenario through well formulated strategies in the short and long term by the year 2000 and 2020 respectively. The Water Resources Act (WRA) 2004 vested rights and control of all water, surface and underground including all water-course affecting more than one state on the federal government through the Federal Ministry of Water Resources.

There have been several policies and strategies in the management of water resources in Nigeria which has not translated to the needed social, economic and environmental goals of sustainable development. The current approach to water resources management has been inadequate. This is manifested in 70 million Nigerians without access to potable water supply and food insecurity. 194,000 Nigerian children under the age of 5 die annually due to water borne diseases [1] . Nigeria also lags behind in the sustainable water resources management which affects water dependent sectors. From irrigated agriculture to power generation, industrial and domestic applications, the inefficiency of the resources management are felt. The Federal Government capital expenditure profile on water infrastructure was N782 billion between the year 2000 to 2012 [3] with no corresponding improvement in the sector.

Okeola and Sule, [4] used analytical hierarchy process (AHP) to study the urban water supply scheme in Offa, Kwara State, Nigeria. The Offa water work was studied with the model to assist its future operation and the design of alternative system improvement works. The study introduces three management options that are formulated on the prevailing nation's water supply sector and foreign countries models. The three options are respectively: Public Ownership and Operation; Public Ownership and Private Operation; and Private Ownership and Operation. The result shows that the stakeholders have chosen option 1 i.e. the Public Ownership and Operation as most contributing to sustainable operation of the urban water supply service delivery under scrutiny of environmental, economical, technical, institutional, and socio-cultural criteria. The choice of option 1 reaffirms the stakeholder opinion in the earlier general survey that they did not want government to abandon their responsibility in the water sector. The ranking of the choice is based on the compromise solutions according to the performance of the system under various alternatives that are considered and the preference of the decision makers (DMs) or stakeholders on Performance Indicators (PI) adopted.

Schardong and Simonovic, [5] in their report presented an application of evolutionary algorithms to multi- objective analysis of water resources systems and their integration into a decision support system (DSS) which are the AcquaNet and ModSim which took the advantage of two multi-objective evolutionary algorithms. The algorithms are developed in the form of DSS which enables generalized multi-objective analysis with a focus on water resources systems. The possibilities of application of the developed DSS and multi-objective evolutionary algorithms are initially exploited by multi-objective analysis of a hydrological rainfall-runoff model Smap. The analysis was then extended to a complex water resources system, the Cantareira System that provides water supply for nearly half of the Sao Paulo Metropolitan Region (SPMR) in Brazil. The analysis was done by comparing two pairs of objective functions: minimization of demand deficits versus minimization of pumping cost and minimization of demand deficits versus minimization of the deviation from the water quality standards. The results show that the multi-objective evolutionary algorithms are suitable for application to integrated water resources management and represent a good alternative to the “classical” methods. This is because satisfactory results were produced and non-dominated solutions were found without much difficulty. They concluded that DSS developed in this study represent important tools for managers and decision makers for management in complex systems considering multiple uses and explore potential solutions to conflicts arising from water use.

Yilmaz and Harmancioglu [6] applied the water resource management model that facilitates indicator-based decisions, with respect to environmental, social and economic dimensions in a multiple criteria perspective for Gediz River Basin in Turkey. The basic input for the model was the quantity of surface water allocated for irrigation purposes. The model was applied under 3 different hydro-meteorological scenarios that reflect baseline as well as better and worse conditions of water supply and demand. This was not only to reach a comprehensive assessment of the water budget in the Gediz Basin but also to evaluate the impacts of proposed management alternatives under different conditions. The Water Evaluation and Planning (WEAP) software was used as simulation and evaluation tool to assess the performance of possible management alternatives. The study delineated the best management alternatives on the basis of three different multi-criteria decision making methods: simple additive weighting (SAW), compromise programming (CP) and technique for order preference by similarity to ideal solution (TOPSIS). The results show that the decision on best alternative was basically independent of the MCDM method used but slightly sensitive to the weights assigned to the criteria as well as the data used in the analyses. It was established that the developed methodology is a valuable tool for the assessment of water resources systems which illustrates an efficient implementation of an integrated water resources management approach for the basin.

RIAM was successfully used to prioritize water resources management problems in Ghana according to the most urgent attention. The priority list was validated and accepted to be the true reflection of the situation at a national workshop in which experts and representatives from water agencies, donor agencies, university faculties and departments, research institutes, private institutions and organizations including Non-Governmental Organizations (NGOs) participated [7] . The study revealed that RIAM can also be used in a developing country like Ghana and is a very useful tool in prioritization process as applied in the exercise.

This study aimed at reviewing and evaluating the status of water resources management in the north central Nigeria, identifying its challenges and prioritizing them to know which requires the most urgent attention and intervention using the Rapid Impact Assessment Matrix (RIAM). This was done through consultation with stakeholders in the sector with vast experiences and understanding of the resources of the region. The region is important considering the role it plays in the water resources of the country.

2. MATERIALS AND METHODS

The methodology employed in this study involves four main stages which are illustrated in the flowchart (Figure 1). Information and data were collected from government documents and gazettes from water related ministries, agencies and organizations. Information was also obtained through discussions with stakeholders. This information aids the design of the questionnaire which was administered. The data obtained from the questionnaires were reduced to capture the most key information needed for RIAM. Subsequently, the data are subjected to RIAM model (coded in excel 2010) for analysis. The outputs of the model are presented graphically in histograms.

An organized workshop consisting of experts and stakeholders in the field of water resources is required for the study. However due to limited fund and logistics, the use of well-structured questionnaire was adopted for these audience. The questionnaire is purposely designed for the stakeholder group in Table 1. A pilot study was first carried out using informal discussions with representatives from the groups. This gave insights on the issues of challenges of management of water resources in their perspectives. Out of the 50 questionnaires that were administered, 34 were completed and retrieved. This is a response rate of 68%. The stakeholders are experts and policy makers in the field of water resources who are spread across different water related organizations in the study area.

Figure 1. Flow chart showing the stages of the methodology employed.

Table 1. Stakeholder groups.

(1) Federal Ministry of Water Resources
(2) State Ministries of Water Resources, Agriculture and Environment
(3) State Water Corporations
(4) Lower Niger River Basin Development Authority
(5) Higher institutions
(6) Water industries in the private sector.

In using the RIAM the following assumptions were considered [7] :

1) Water resources management problems are considered to impact negatively.

2) Positive impacts of water resources management issues do not pose any threat and therefore not considered relevant.

3) The importance of a problem could be localized or extensive i.e. affect the immediate environment only or can spread beyond.

The RIAM method is based on a standard definition of the important assessment criteria, as well as the means by which semi-quantitative values for each of these criteria can be collated to provide an accurate and independent score for each condition [8] . The assessment criteria are divided into two categories:

1) Group (A) criteria

Criteria that are of importance to the condition and which can individually change the score obtained.

2) Group (B) criteria

Criteria that are of value to the situation but individually should not be capable of changing the score obtained.

The value of each of these groups of criteria is determined by the use of a series of simple formulae which allow the scores for the individual components to be determined on a defined basis. The values of score in group (A) criteria are multiplied together while those of group (B) criteria are summed up. The final assessment score for the condition is obtain by multiplying the result of group (A) by the result of group (B) indicated in Equations (1), (2) and (3) respectively.

$$(A1)\times(A2)=AT \tag{1}$$

$$(B1)+(B2)+(B3)=BT \tag{2}$$

$$(AT)\times(BT)=ES \tag{3}$$

where,

(A1) and (A2) are individual criteria scores that are of importance to the condition (group A) and which can individually change the score obtained. (B1) to (B3) are the individual criteria scores that are of value to the situation (group B) but individually should not be capable of changing the score obtained.

AT is the result of multiplication of all (A) scores;

BT is the result of summation of all (B) scores; and

ES is the assessment score for the condition.

2.1. Study Area

The north central Nigeria (Figure 2) consists of six states, namely: Benue, Kogi, Kwara, Nassarawa, Niger, Plateau, and the Federal Capital Territory (FCT). The region has a population of 20.3 million (2006 estimate) and the main economic activity is agriculture. Other activities include fishing, mining, and businesses. The most significant feature of the region in terms of water resources is the Niger Basin with an area of 584,193 km^2 within the country which is 63 percent of the total area of the country and covers a large area in north central region [9] . There has been an increase in the social and economic activities in this region in the past decades due to the location of the FCT which has continue to attract people thus adding pressure to the existing water infrastructure.

2.2. Rapid Impact Assessment Matrix (RIAM)

The rapid impact assessment matrix (RIAM) is one of the frontiers tool commonly employed in the study of environmental impact assessment (EIA). RIAM uses a structured matrix to allow for judgments both subjective and those based on quantitative data to be made on a like-by-like basis, and provides a transparent and permanent record of the judgments made [8] . The concept of RIAM was developed to improve the existing key shortcomings of the EIA methods. According to Kankam-Yeboah et al., [7] , the simple and structured form of RIAM allows reanalysis and in-depth analysis of selected components in a rapid and accurate manner. This flexibility makes the method powerful tool for both executing and evaluating impact assessments. RIAM is based on the knowledge that certain specific criteria are common to all impact assessments and by scaling these criteria, it becomes possible to record the values of the assessments made. It was adopted for study to prioritize the water resources management to know which problems require the most urgent attention and mitigation measures particularly in the eyes of stakeholders who at the receiving end of policy implementation.

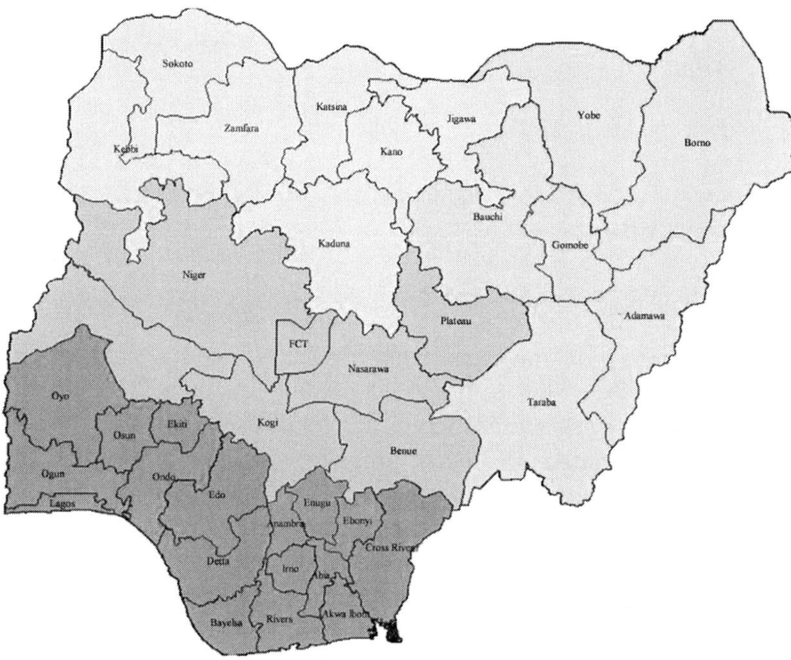

Figure 2. Map of Nigeria showing the north central zone. ▭: North Central Nigeria

2.2.1. Water Resources Management Problems

The following eleven problems were identified by the stakeholders to be most severe in the north central region in order of severity.

1) Bad governance.

2) Inadequate data for planning.

3) Misappropriation of available funds.

4) Pollution.

5) Poor data collection and banking.

6) Limited groundwater in some areas.

7) Contaminated source of water.

8) Flood.

9) Indiscriminate groundwater exploitation.

10) Inadequate funds for further agricultural, hydroelectric, navigational and industrial development.

11) Lack of operational sustainability.

Each of these problems was then evaluated by the stakeholders by assigning scores in Table 2. The RIAM model (Equations (1)-(3)) was then applied to get the Assessment Scores (ES) for each component. The average of the ES calculated for each problem was taken along with the corresponding Range Band selected in Table 3. The scoring was done by experts from water related disciplines. They have acquired vast experiences over the years working in water related discipline. They also have firsthand information on the efficient management of water resources in the region which qualifies them for the scoring in this technique.

2.2.2. Impact Indicators

This study adopted the following impact indicators which are among the most relevant and widely used indicators for the assessment and analysis of water management problems [7] [10] :

1) Environmental quality.

2) Degree of tolerance to the public.

3) Political and legal feasibility.

4) Equity and gender issue.

5) Economic growth.

Table 2. Assessment criteria.

Criteria	Scale	Description
A1: Importance of condition	4	Important to national/international interest
	3	Important to regional/national interest
	2	Important to areas immediately outside the local condition
	1	Important only to local condition
	0	No importance
A2: Magnitude of change/effect	+3	Major positive benefits
	+2	Significant improvement in status quo
	+1	Improvement in status quo
	0	No change/status quo
	−1	Negative change in status quo
	−2	Significant negative change
	−3	major change
B1: Permanence	1	No change/not applicable
	2	Temporary
	3	Permanent
	1	No change/not applicable
	2	Reversible
	3	Irreversible
B3: Cumulative	1	No change/not applicable
	2	Non-cumulative/single
	3	Cumulative/synergistic

Source: [8].

Table 3. Conversion of environmental scores to range bands.

Environmental score	Range bands	Description of range bands
−72 to +108	+E	Major positive change/impacts
+36 to +71	+D	Significant positive change/impacts
+19 to +35	+C	Moderately positive change/impacts
+10 to +18	+B	Positive change/impacts
+1 to +9	+A	Slightly positive change/impacts
0	N	No change/status quo/not applicable
−1 to −9	−A	Slightly negative change/impacts
−10 to −18	−B	Negative change/impacts
−19 to −35	−C	Moderately negative change/impacts
−36 to −71	−D	Significantly negative change/impacts
−72 to −108	−E	Major negative change/impacts

Source: [8].

6) Technical feasibility.

7) Health care delivery.

8) Safe water supply and consumption.

9) Biodiversity/ecosystem.

10) Water quality enhancement.

11) Cost of option.

12) Energy supply.

13) Security and reliability of water supply.

These impact indicators were subsequently grouped into four components (Table 4).

3. RESULT AND DISCUSSIONS

The histogram of the RIAM result provides a comparative picture of the negative impacts of all the problems assessed shown Figures 3-13. The Range Bands are represented on the horizontal scale of the histograms. N stands for neutral and the negative alphabets shows the negative impacts of the component. The vertical scale indicates the scores. PC represents physical/chemical component, BE represents biological/ecological component, SC represents sociological/cultural component and EO represents economical/operational component.

The most severe problem is that with the highest number of negative scores. In this study "Inadequate funds for further agricultural, hydroelectric, navigation and industrial development" recorded the score of 14 under (−D) which is the maximum score obtained for all the criteria in the component and 3 under (−C). This is followed closely by "Poor data collection and banking" with a score of 13 under (−D) and 4 under (−C). However," Indiscriminate groundwater exploitation", "Bad governance" and "inadequate data for planning" tie with the score of 11 under (−D) and 6 under (−C). The implication is that the three problems are considered to have the same level of urgency as regards the intervention measures they require. "Flood" is the next with the score of 11 recorded under (−D), 5 under (–C) and 1 under (–B) which is followed closely by "Misappropriation of funds" with the score 10 under (–D), 6 under (–C) and 1 under (–B). "Pollution" records 9 under (–D) and 8 under (–C); "Contaminated sources of water" records 9 under (–D), 6 under (–C) and 2 under (–B); "Lack of operational sustainability" records 8 under (–D) and 9 under (–C) and the problem requiring the least attention on the list is "Limited groundwater in some areas" with the score 7 recorded under (–D) and 10 under (–C).

The overall representation of all components is shown in Figure 14. The Figure indicates a total of 114 scores registered under the –D (significantly negative impacts), 69 scores registered for –C (moderately negative impacts) and a score of 4 for –B (negative impacts). A major characteristic of the RIAM model is that the most significant impact (–D) is considered as the impact that should receive the most urgent attention in terms of mitigation and intervention measures before the next significant impact (–C). Consequently, a problem registering a single –D takes precedence over a problem which registers say twice or more –C in terms of which of them requires the most urgent attention and intervention. The significantly negative impacts having registered the highest scores indicate that majority of the evaluated problems are considered to require urgent attention and mitigation measures. However, the order in which they are to be selected for quick intervention is provided by the RIAM in the priority list inTable 5.

The result of the prioritization of water resources management problems using the RIAM technique is summarized in Table 5. "Inadequate funds for further agricultural, hydroelectric, navigation and industrial development" recorded the most negative impacts with a score of –14 registered under (–D) and –3 under (–C). This result supported the work of [11] - [13] in their various study of application of severity index in matrix order (SIMO) model in empirical ranking where "insufficient funding for infrastructures" was listed as one of the critical factors to infrastructural service delivery failures in the Nigeria.

The results obtained from RIAM tool was compared with the stakeholders' conventional ranking of the different categories of water resources management problems and also that of individual ranking of problems. While individual problem ranking was different from that of RIAM, they both agree with the ranking of categories of water resources problems. This could be as a result of the usage of questionnaire instead of an organized workshop which facilitates insightful thoughts and inclusive discourse. The questionnaires completion at different periods with extended interval of time easily results in discontinuity in their critical thinking.

Table 4. Components and description.

S/No.	Components	Description
1.	Physical/chemical	This group covers all physical and chemical aspects of the water resources management, including non-renewable natural resources and degradation of the physical environment through pollution.
2.	Biological/ecological	This group includes all biological aspects of the water resources covers management, including renewable natural resources, conservation of biodiversity, species interaction and pollution of biosphere.
3.	Sociological/cultural	This group encompasses human aspects of the water resources management, including social issues affecting individuals and community; together with cultural aspects including conservation of heritage and human development.
4.	Economic/operation	This group is used to qualitatively identify the economic consequences of change in water resources management, both temporary and permanent.

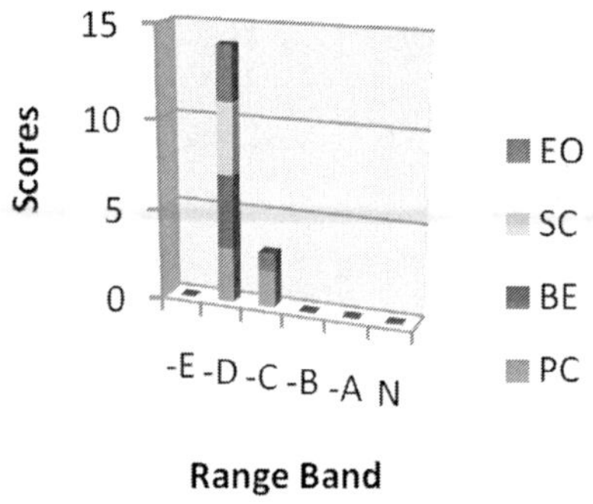

Figure 3. RIAM output for "inadequate, funds for further agricultural, hydro-electric, navigational and industrial development".

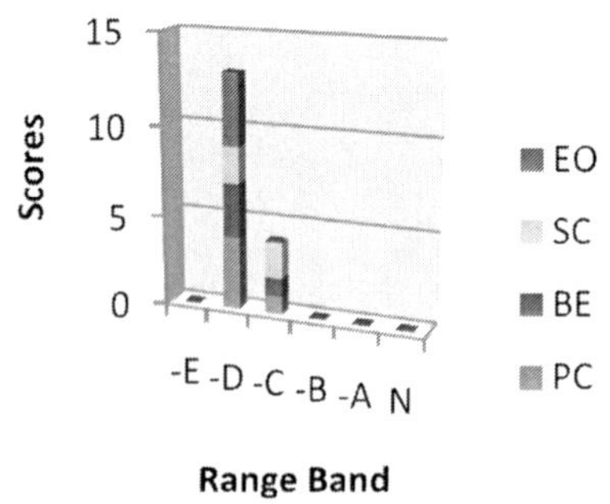

Figure 4. RIAM output for "poor data collection and Banking".

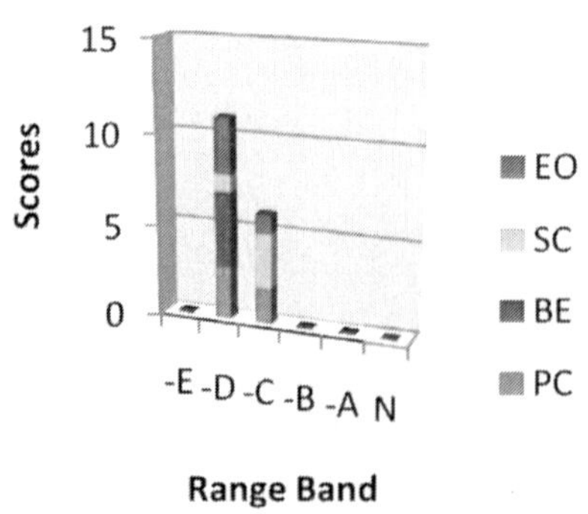

Figure 5. RIAM output for "indiscriminate groundwater exploitation".

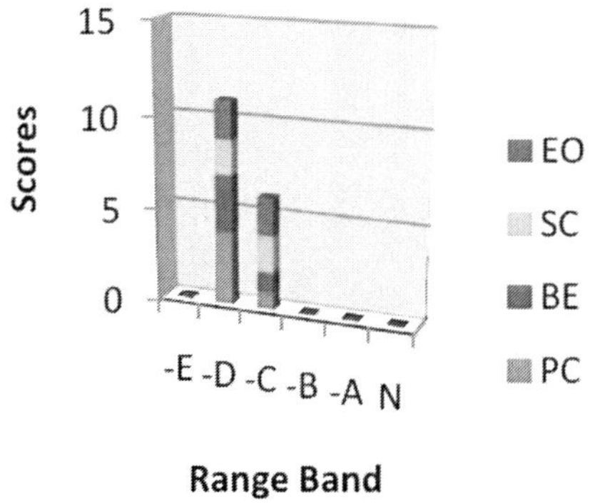

Figure 6. RIAM output for “bad governance”.

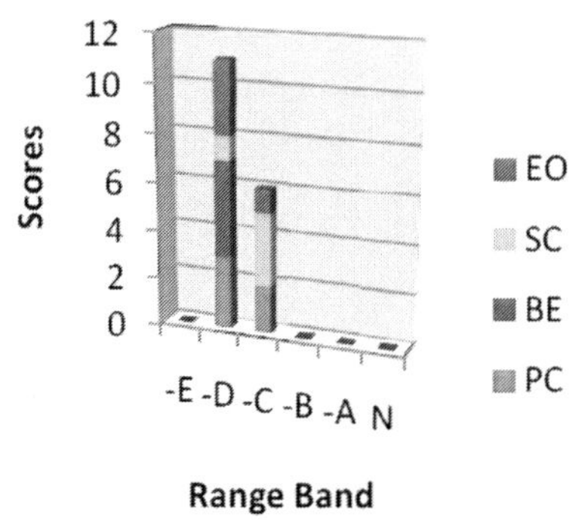

Figure 7. RIAM output for “inadequate data for planning”.

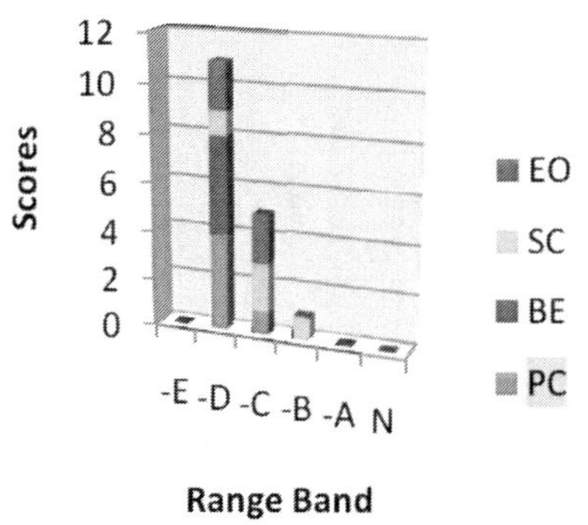

Figure 8. RIAM output for “flood”.

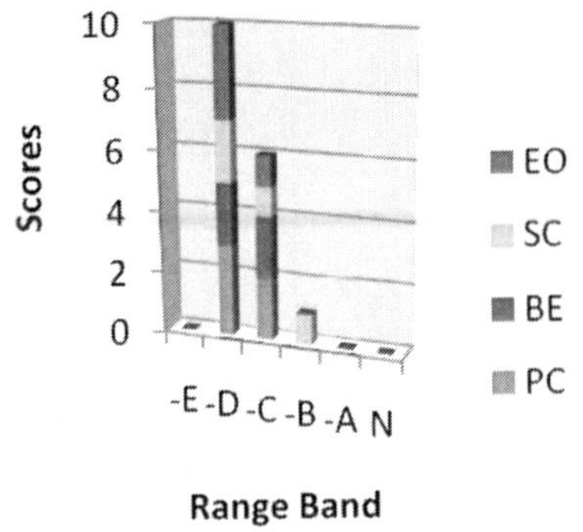

Figure 9. RIAM output for "misappropriation of available funds".

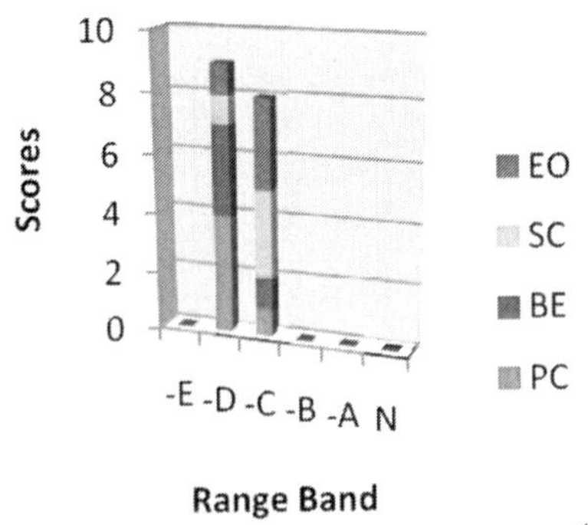

Figure 10. RIAM output for "pollution".

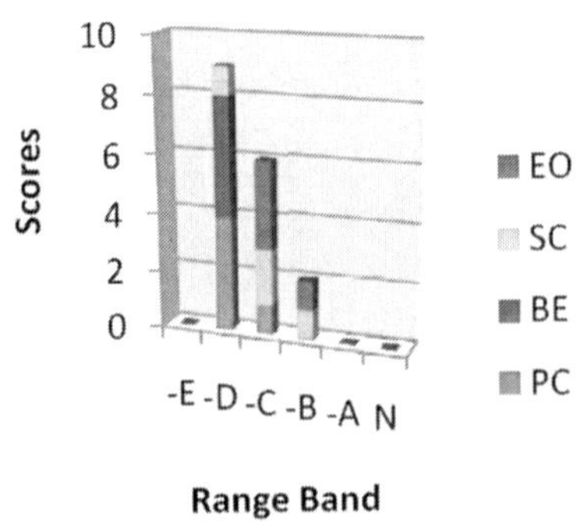

Figure 11. RIAM output for "Contaminated sources of water".

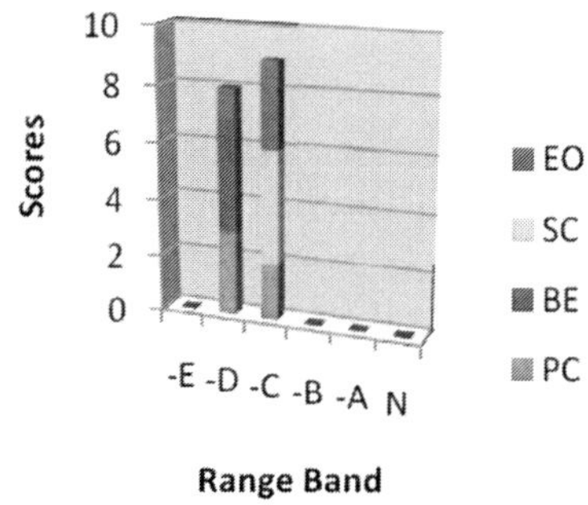

Figure 12. RIAM output for "lack of operational sustainability".

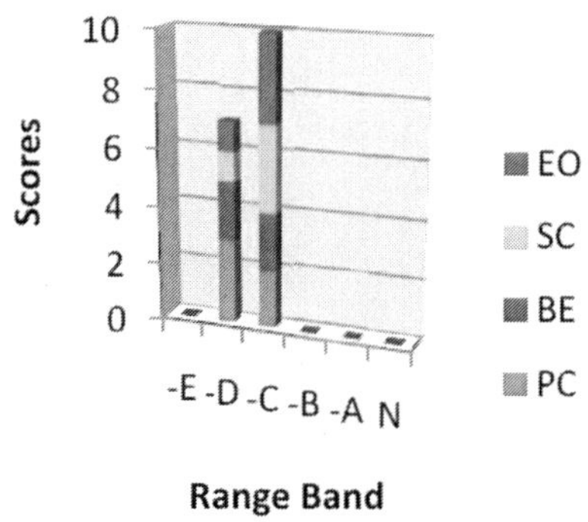

Figure 13. RIAM output for "limited groundwater in some areas".

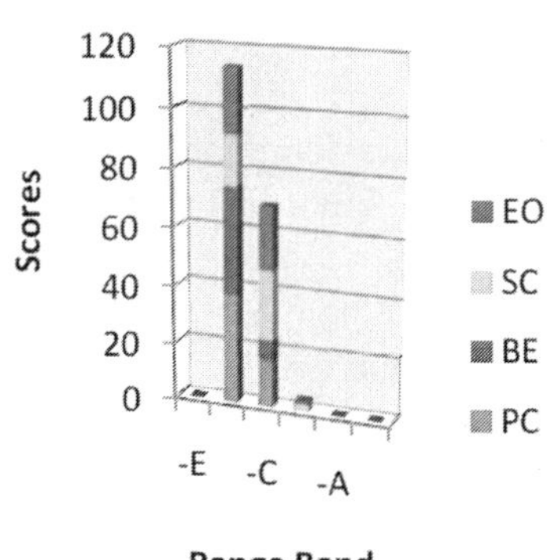

Figure 14. Overall representation of the RIAM output.

4. CONCLUSIONS

The RIAM technique has been successfully applied to the north central water resources management. It has made a key contribution to the prioritization of water resources management by providing insight into the most urgent problem that requires immediate intervention for the zone at least according to policy implementers.

Table 5. Result of the prioritization of the water resources management problems.

No.	Water resources management problems	–E	–D	–C	–B	–A	N
1	Inadequate funds for further agric. Hydroelectric, navigation and industrial development		14	3			
2	Poor data collection and banking		13	4			
3	Indiscriminate groundwater exploitation		11	6			
4	Bad governance		11	6			
5	Inadequate data for planning		11	6			
6	Flood		11	5	1		
7	Misappropriation of the available funds		10	6	1		
8	Pollution		9	8			
9	Contaminated sources of water		9	6	2		
10	Lack of operational sustainability		8	9			
11	Limited groundwater in some areas		7	10			
	Total		**114**	**69**	**4**		

Eleven problems were analyzed by RIAM technique based on the physical/chemical, biological/ecological, social/cultural and economic/operational factors using several impact indicators. The RIAM identified 1) "inadequate funds for further agricultural, hydroelectric, navigation and industrial development" as the number 1 challenge demanding attention and 2) "poor data collection and banking" as the number 2. The RIAM obtained result constitutes a vital and insightful guide for decision makers in articulating effective sustainable policy and reforms in the management of the region's water resources.

It is recommended that the key stakeholders (FMWR and LNRDA) set in motion a strategy to address the challenges that will have widespread citizenry appreciation. This can be done through collaborative and interdisciplinary research using this study as baseline. It is also imperative to extend this application to other geopolitical zone of the country.

REFERENCES

1. Federal Government of Nigeria (2011) Executive Summary of the Nigeria Water Sector Roadmap. Federal Ministry of Water Resources, Abuja.

2. Federal Ministry of Water Resources (2014) http://www.waterresources.gov.ng/about/core-functions

3. Ali, K.A. (2012) Development of Water Supply Infrastructure in Nigeria: Challenges and Prospects. Nigeria Society of Engineers October Lecture, Abuja

4. Okeola, O.G. and Sule, B.F. (2011) Evaluation of Management Alternatives for Urban Water Supply System Using Multicriteria Decision Analysis. Journal of King Saud University—Engineering Science, 24, 19-24.

5. Schardong, A. and Simonovic, S.P. (2011) Multi-Objective Evolutionary Algorithms of Water Resources Management. Water Resources Research Report. Report No. 078, Department of Civil and Environmental Engineering, University of Western Ontario, Canada

6. Yilmaz, B. and Harmanacioglu (2010) Multi-Criteria Decision Making for Water Resources Management: A Case Study of Gediz River Basin, Turkey. Water SA, 36, 563-576.

7. Kankam-Yeboah, K., Asare, E.B., Gyau-Boakye, P. and Nishigaki, M. (2005) Rapid Impact Assessment Matrix (RIAM)—An Analytical Tool in the Prioritization of Water Resources Management Problems in Ghana. Journal of the Faculty of Environmental Science and Technology, Okayama University, 10, 75-81.

8. Pastakia, C.M.R. and Jensen, A. (1998) The Rapid Impact Assessment Matrix (RIAM) for Environmental Impact Assessment. Environmental Impact Assessment Review, 18, 461-462.

9. Food and Agriculture Organization of the United Nations (2005) AQUASTAT—FAO's Information System on Water and Agriculture. http://www.fao.org/waicent/faoinfo/agricult/agl/aglw/aquastat/countries/nigeria/index.stm

10. Hajkowicz, S. and Collins, K. (2007) A Review of Multiple Criteria Analysis for Water Resources Planning and Management. Water Resources Management, 21, 1553-1566.

11. Okeola, O.G. and Salami, A.W. (2012) A Pragmatic Approach to the Nigeria's Engineering Infrastructure Dilemma. Epistemics in Science, Engineering and Technology, 2, 55-61.

12. Adenikinju, A. (2005) Analysis of the Cost of Infrastructure Failures in Developing Economy: The Case Study of Electricity Sector in Nigeria. AERC Research Paper 148, African Economic Research Consortium, Nairobi.

13. Omoregie, A., Ebohon, O.J. and Radford, D. (2011) Modeling in Ranking Procedures: A Case Study: Infrastructure Failure in Nigeria. School of Architecture, De Montfort University, Leceister. http://www.irbdirect.de/daten/iconda/cib10618.pdf

Chapter 8

Coastal Reservoir Strategy and Its Applications

Shuqing Yang[1], Jianli Liu[1], Pengzhi Lin[2] and Changbo Jiang[3]

[1] School of Civil, Mining& Environmental Engineering, University of Wollongong, Australia

[2] State Key Lab. of Hydraulics and Mountain River Engineering, Sichuan University, China

[3] College of Hydraulic Engineering, Changsha University of Science and Technology, China

1. INTRODUCTION

While the world's population tripled in the 20th century, the use of renewable water resources has grown six-fold [1]. It is estimated that the world population will enlarge by another 40 to 50 % in the following fifty years. The demand for water will be increasing resulted by the population growth combined with industrialization and urbanization, which will have serious consequences on the environment. According to WHO/UNICEF Joint Monitoring Programme (JMP) (2012 Update), 780 million people lack access to an improved water source; approximately one in nine people [2]. Water stress causes deterioration of fresh water resources in terms of quantity (aquifer over-exploitation, dry rivers, etc.) and quality (eutrophication, organic matter pollution, saline intrusion, etc.). In the Developing World, women and children walk miles to get water. The UN estimates that the average is 40 pounds of water carried 4 miles (18 kg-6 km). This takes hours, people can't attend school/work, deforms the spine and can leave women vulnerable to assault [3]. Figure 1 showed the state of water shortages based on synthetic evaluation of water management using for agriculture in 2007.

As about 66% of the world population would be confronted with water-shortage by 2025 [5]; the water from aquifers, which provide water for one-third of the world's population, are being used out before nature can complement them [6]. Water scarcity is already a focus of attention all over the world [7]. For example, the Southwest and Midwest areas of the USA and Australia are vulnerable to water scarcity [8]. In Australia, water allocations for irrigation have caused a conspicuous decline in rainfall and runoff in the past decades [9]. Figure 2 shows drinking water as the substance to transform in a social life. Many indigenous peoples also depend very much on natural resources and live in ecosystems especially vulnerable to the effects of climate change, such as small island developing states, arctic regions and high altitudes [10].

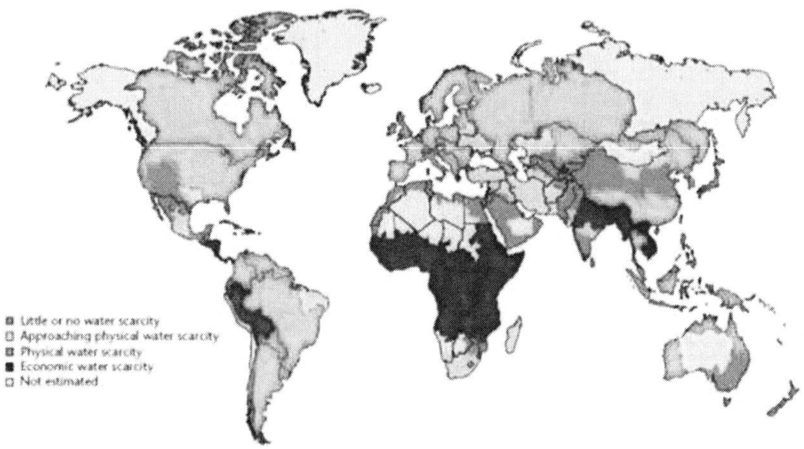

Figure 1. Water scarcity based on synthetic evaluation of water management using for Agriculture in 2007 [4]

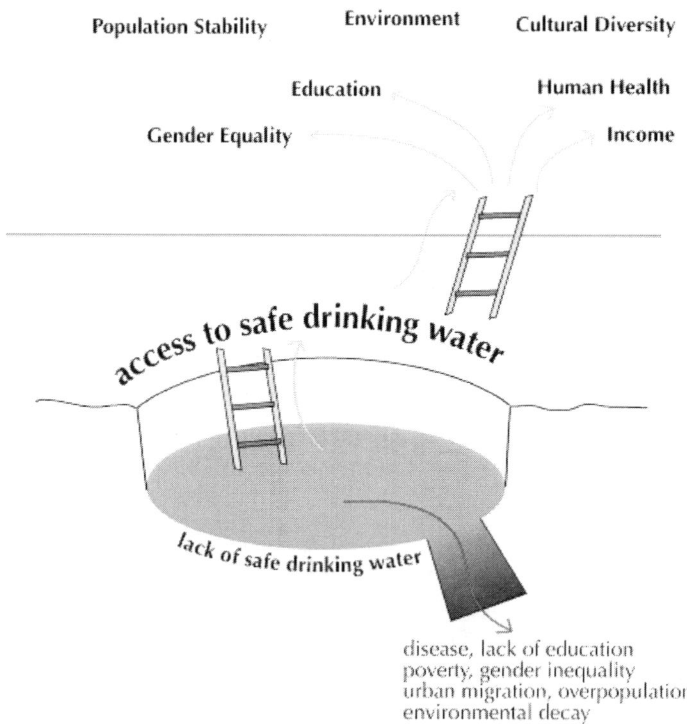

Figure 2. Drinking water as the substance to transform a life [11]

The shortage of fresh water reserves is a world-wide problem nowadays, which can lead to a global crisis. Firstly, it can cause agricultural crisis. 70–85 percent of water use is for agriculture, and it is assessed that the most imperilling agricultural growth is 20 percent of global grain production won't have enough water to irrigate in the future. Connected with spatial and temporal variations, water is not always available, which means there is scarcer water particularly for large concentrations of people consumption, enterprises and some other uses. Furthermore, it leads to environmental crisis. In Water and Nature, the more critical statement has been come out that the increasing of using water resource can reduce the amount of water available for industrial and agricultural development, more seriously that can exert a far-reaching influence on aquatic ecosystems and their dependent species. Environmental balances are disturbed and cannot play their roles like before [12]. More than 260 river basins are shared by two or more countries in the world. The lack of water can lead to transboundary tensions. As time passes, a series of conflicts come out, which increases regional instability, such as The Parana La Plata, the Aral Sea, the Jordan and the Danube.

Today, approximately 3 billion people — about half of the world's population — live within 200 kilometers of a coastline. By 2025,that figure is likely to double. In many countries, populations in coastal areas are growing faster than those in non-coastal areas. Take the example of China, the world's most populous nation. Of China's 1.2 billion people, close to 60% live in 12 coastal provinces, along the Yangtze River valley, and in two coastal municipalities — Shanghai and Tianjin. Along China's 18,000 kilometers of continental coastline, population densities average between 110 and 1,600 per square kilometer. In some coastal cities such as Shanghai, China's largest with 17 million inhabitants, population densities average over 2,000 per square kilometer. From 1990 to 2015, people flock to coastal cities continuously, especially in the past ten years, the number of migrated population is more and more bigger. Figure 3 [13] shows the population changes by 5 years (from 1990-2015) in Low elevation coastal zones (LECZ is a continuous strip along coastal areas whose altitude is less than 10m). In the United States, over 50 percent of the nation's population lives in only less than 20 percent of the U.S. coastal area (excluding Alaska) in 2010 [14]. Australia has experienced the same situation. Australia's coastal population has been growing faster than the population of the rest of the country for some time [15] (see Figure 4) and is expected to increase by another one million people over the next 15 years [16]. Up to 50 percent of the population in northern Africa and Bangladesh lives in coastal areas; along the Nile Delta, the population density reaches 500 to 1,000 people per square kilometre [17]. The high concentration of people in coastal regions has produced many economic benefits. But the combined effects of booming population growth and economic and technological development have aggravated the need for water.

Ban Ki-moon warns: "A shortage of water resources could spell increased conflicts in the future. Population growth will make the problem worse. So will the climate change. As the global economy grows, so will its thirst. Many more conflicts lie just over the horizon." [4]

With the current state of affairs, correct measures are needed to be taken to avoid the crisis to be worsening. There is an increasing awareness that our freshwater resources are limited and need to be noticed quantity and quality. "The time for solutions" is put forward in The 6th World Water Forum. This study summarized the overall situation of water solutions in the past and proposed a new waters strategy-coastal reservoir.

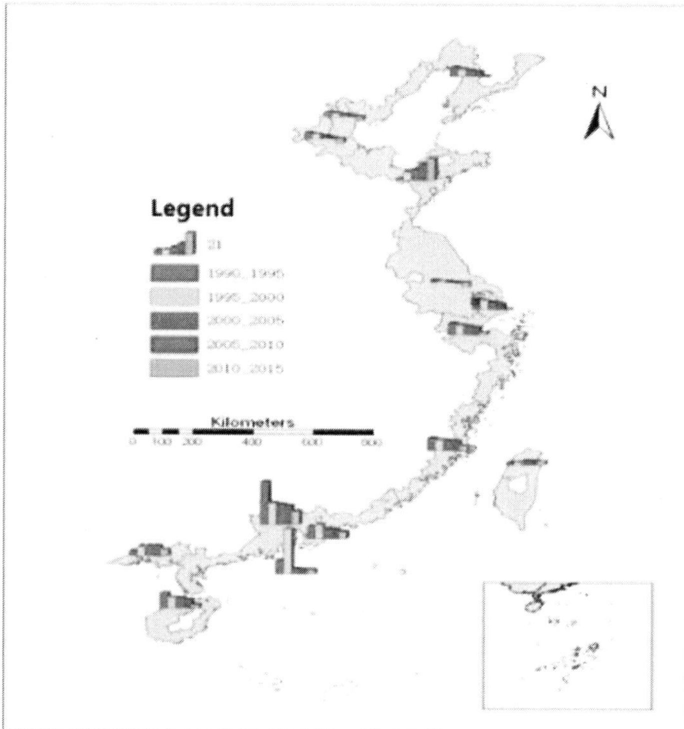

Figure 3. Coastal Population in China

2. CURRENT WATER SOLUTIONS

There are mainly several kinds of water supply options in the world, for example, groundwater, on-land reservoir, desalination of seawater, reuse of wastewater, Diversion of water from a remote source. They have their own characters.

Groundwater generally has higher chemical, physical and biological quality than surface waters. This higher quality means that groundwater usually requires lower levels of treatment, and the treatment process is therefore cheaper. But Groundwater is harder to locate than surface water. The location of groundwater may require the use of sophisticated equipment. But groundwater resources in the world are almost fully developed, and in some cases they are already considered overdeveloped. The groundwater sources that could potentially be developed in the world are so small that it would not be economically viable to do so. Therefore, the next preferable water supply source must be considered.

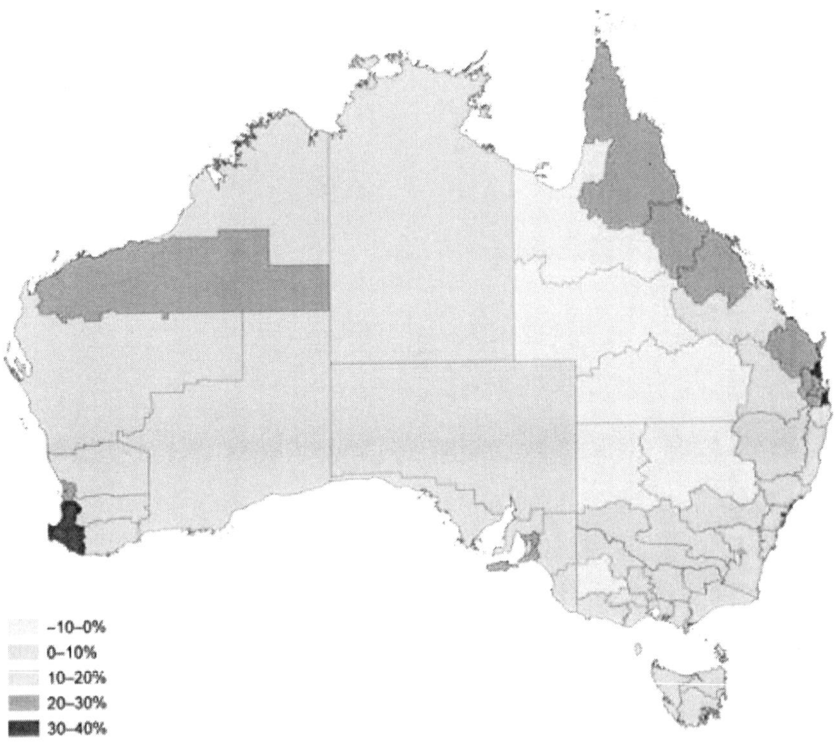

Figure 4. Coastal Population in Australia

The water quality in on-land dams is higher than that of the water from all the alternative sources except groundwater. But surface waters are more susceptible to pollution than groundwater. They are difficult to protect from contamination (such as waterborne diseases and chemicals that enter from surface runoff and upstream discharges). The water quality is also not as consistent. For instance, turbidity and temperature fluctuate, often depending on the amount of precipitation. Surface water requires more treatment than groundwater and is therefore more expensive to utilise. This water supply source relies on rainfall. There are also usually a limited number of dam sites available because the construction of dams requires the correct combination of topography and geology. This is due to a lack of appropriate sites for the construction of dams. For example in Australia, after two centuries of development, there are almost no new dam sites left with good combination of topography and geology. Soil erosion also keeps reducing the storage capacity of the world's on-land reservoirs by more than 1% annually. For example, on-land reservoirs were subjected to very high siltation rates which are comparable to overseas extreme siltation rates in Australia [18]. In the future, soil erosion and reservoir sedimentation rates would be accelerated due to the severity of storms and rains as a result of global warming [10]. This means that almost all existing on-land reservoirs may lose its capacity in about 100 years. Therefore, the next water source option should be considered.

The first desalination plant by the multi-stage flash (MSF) evaporation process installed in 1964 in Lanzarote(Canary Islands) to over 700 desalination plants existing nowadays [19].Desalination technologies account for a worldwide production capacity of 24.5 million m^3/day [20], the cost decreased to \$0.50-0.80/m^3 desalinated water and even to 0.20-0.35 \$/m^3 for treatment of brackish water [21]. However, they have potential negative impacts [22]. These are mainly attributed to the concentrate and chemical discharges, which may impair coastal water quality and affect marine life, and air pollutant emissions attributed to the energy demand of the processes. A key concern of desalination plants are the concentrate and chemical discharges to the marine environment, which may have adverse effects on water and sediment quality, impair marine life and the functioning and intactness of coastal ecosystems [22]. WHO [23] provided a general overview on the composition and effects of the waste discharges in their guidance document. Lattemann [22, 24] and MEDRC [25] discussed details of the negative impacts. Reverse osmosis (RO), a commonly used desalination technology, is significantly more expensive than the standard treatment of fresh- water for potable use [20]. Furthermore, desalination technologies have the risk of oil pollution [26], Most of the noise is produced by the high-pressure pumps and by the turbines used for energy restoration [27].Desalination of seawater does not rely on rainfall (it is a climate resilient water source). Desalination plants are located on the coast, and as mentioned earlier this is where the majority of the water demand is expected to be located. But the treatment process would have high energy requirements as well as high operating and maintenance costs because of the high concentrations of contaminants that would have to be removed. For instance, seawater is usually 30 times more saline than the sewage effluent that would need to be treated to produce PRW (QWC, 2008). The concentrate stream produced by the desalination process can have detrimental environmental impacts. This could be a significant problem if the desalination plants were located on the coast, because the discharge would be released into a place, which is maybe a very sensitive marine environment. Therefore, other water supply options need to be considered.

Diversion of water from a remote source does not depend on local rainfall or local water quality. But this water source would have very high construction costs. For instance, the government of Queensland in Australia has considered diverting water supplies from North Eastern NSW, but this was found to be not economically viable. There were also numerous social, environmental and interstate issues that were considered to be insurmountable (QWC, 2008). Water could potentially be diverted from Northern Queensland. However, this would involve very high construction costs and should only be considered if no alternative water sources are available locally.

Reuse of wastewater does not depend directly on rainfall. But the levels of contaminants are higher than for the previous water supply options. Therefore, the cost of operating the treatment process would be higher. It is a common practice to discharge untreated sewage directly into bodies of water or put onto agricultural land, causing significant health and economic risks [28]. There could also be public perception problems with the reuse of wastewater (this is often referred to as the 'Yuk Factor').

Stormwater harvesting is used in house yards, streets and parks, but there are several shortages. CM Goonrey et al. exam the technical feasibility of using stormwater as an alternative supply source in an existing urban area [29]. Firstly, its storage capacity is very small due to the structural constraints; then its water quality may be not very good as the water comes from densely populated areas, and its cost is also very high as precious land is used to store the storm water.

Now in the world, many people try to find ways to solve water scarcity, however, on the other hand,too much stormwater and flood can't be neglected. The former management policies lead to vast amount of floodwater being discharged into the sea. Drought and lack of water emerge during dry season. It is obvious that new waters storages are in demand in the country. The coastal reservoir is to solve water storage of the next generation. It is a dam in the sea at the mouth of river or near it. It catches water in the bottom of the stream [30]. The building imitates the natural freshwater lake with a sustainable flow of fresh water from the river. Such reservoirs were already used in the Asian countries – Singapore, Hong Kong, China – and they are considered as very successful alternative to traditional dams.

"There is a water crisis today. But the crisis is not about having too little water to satisfy our needs. It is a crisis of managing water so badly that billions of people - and the environment - suffer badly." ——World Water Vision Report

3. COASTAL RESERVOIR

A Coastal Reservoir is a freshwater reservoir located in the sea at the mouth of a river with a sustainable annual river flow [31]. All a costal reservoir needs to be effective is an impermeable barrier between the fresh river water and the salty sea water. Yang outlined three guidelines for the successful construction of a coastal reservoir. The first guideline is separation, meaning the successful separation of clean river from polluted water and salt water. Next is protection, meaning the protection of collected fresh water against polluted river water and external pollution. Last is prevention, meaning the successfully prevention of salt water intrusion into the stored fresh water; weather that be through permeability or large tidal events.

Compared with water from seawater desalination processes, catchment runoff is a natural resource, which is cost-saving and water quality as the basic guarantee, is more closed to drinking water. Different from the on-land catchment runoff proposal, coastal reservoir harvests the catchment runoff in the sea, its water sources from a river and the reservoir has the potential to catch all runoff from a catchment. The coastal reservoir can be classified into various categories, in terms of location, barrage, and water quality etc. Existing freshwater lakes or lagoons on the shore can be regarded as special or natural coastal reservoirs. The main differences between coastal reservoirs and on-land reservoirs are summarized in Table1.

Table 1 shows when constructed they share many similarities to inland reservoirs but due to the presence of water on either side which have almost the same density the hard barrier does not need to be constructed as strong as an inland reservoir does (material costs are a lot lower).

Table 1. Differences between on-land Reservoirs and Coastal Reservoirs

Item	On-land Reservoir	Coastal Reservoir
Dam-site	Valley(limited area)	Coast(inside/outside river mouth)
Water level	Above sea level	At sea level
Pressure	High pressure	Low pressure but with wave surge
Seepage	By head difference	By density difference
Pollutant	Land based	Land-based and seawater
Land acquisition	High	Low
Water supply	By gravity	By pump

3.1. Existing Coastal Reservoirs

The first coastal reservoir was built in Zuider Zee, Netherlands in 1932, named Ijsselmeer with water area of 1240 km^2. At that time, people mainly enclose sea area to set up land with getting water as a fringe benefit. During the late 20th century and early 21st century, these are predominantly man made instances of coastal reservoirs with accelerated dilution of salt water to form viable storage and catchment of potable water.

The construction of these coastal reservoirs involves forming a dam wall, usually of solid material across the point where a river or lake enters the ocean. Salt water dammed in the reservoir is then pumped out to aid the speed in which water is diluted to form fresh water. As this reservoir is formed at sea level, maximum flow will occur and full use of the catchment inflows feeding the river or lake will be made. Ordinary storage reservoirs that involve river dams also house water from catchment inflows however these reservoirs are commonly located further upstream which allows all water below the dam to be released into the ocean. This type of Coastal Reservoir is being used in Singapore, South Korea, Hong Kong and China. Currently, there exist many coastal reservoirs in the world as listed inTable2.

Table 2. Existing Coastal Reservoirs in the World

Name	Catchment (km²)	Dam length(m)	Capacity (million m³)	Year completed	Country/river
Qingcaosha	66.26	48786	435	2011	China/ Yangtze
Saemanguem		33900	530	2010	South Korea
Sihwa	56.5	12400	323	1994	South Korea
Marina Barrage		350	42.5	2008	Singapore
Chenhang				1992	China/Shanghai
Yu Huan	166	1080	64.1	1998	China/Zhejiang
Baogang				1985	China/Shanghai
Plover Cove	45.9	2000	230	1968	Hong kong

3.2. The Types and Functions Of Coastal Reservoir

The coastal reservoir can be classified into various categories, in terms of location, barrage, and water quality etc. According to the geographical location, it can be divided to estuary reservoir, intertidal reservoir, gulf reservoir. According to the water quality, it can be divided to drinking water reservoir with good quality, freshwater reservoir for agricultural/industrial purpose with moderate quality, sewage reservoir, ballast water reservoir, etc. According to the dam body, it can be divided to concrete dam, earth dam and soft dam reservoirs. And it also can be divided to natural and aritificial, for example, Saemangeum coastal reservoir, South Korea, which is a huge project of the century.

Coastal reservoir can be used to provide water to three main parts, irrigation, industrial and domestic water using. For example, Chenhang Reservoir, which is located in Yangtze River Estuary of China, mainly provides drinking water for north area of Shanghai (see Figure.5). Baogang Reservoir, which is located in Yangtze River Estuary of China, plays an important role in providing industrial water to keep Baoshan Steel going on (see Figure5).

3.3. Plover Cove Reservoir In Hong Kong

Plover Cove Reservoir, located within Plover Cove Country Park, in the northeastern New Territories, is the largest reservoir in Hong Kong in terms of area, and the second-largest in terms of volume. It was the first in the world to construct a lake from an arm of the ocean. Its main dam was one of the largest in the world at the time of its construction, disconnecting Plover Cove from the sea.

Figure 5. Baogang and Chenhang Reservoir (**1** is Baogang Reservoir and **2** is Chenhang Reservoir)

The dam of the reservoir is 28m tall and approximately 2 km long, which was built by layers of sand and gravel. Besides rain from its catchment, it also stores water imported by pipes from Dongjiang. The Bride's Pool flows into the Plover Cove Reservoir. One main dam and three service dams were built to shut the cove off from the sea. The cove was then drained and was converted into a fresh-water lake.

The location was a former cove (bay, as the name suggests) and was a popular hiking site. Construction work commenced in 1960 and was completed in 1968, providing a capacity of 170 million m^3. Work on raising the height of the dams began in 1970. Upon completion in 1973, the reservoir capacity was increased to 230 million m^3 (see Figure 6).

Figure 6. Plover Cove Reservoir in Hong Kong

3.4. Coastal Reservoir In Singapore

Singapore is a tropical coastal city between the lines of longitude E103° 38' and 104°05', the latitude N 1°09' and 1°29' with population of 4 million and area of 680km^2. The mean annual rainfall is 2.4m. The total evapo-transpiration, infiltration, etc. loss is around 1.17m to 1.27m per year. At the end of 2000, total volume of water consumed in Singapore was 455.4 million m^3/year. Half of Singapore's water supply is imported from Malaysia across Johor strait through a Causeway, and the other half of water demand comes from its own reservoirs.

Due to its rapid economic and population growth in the past decades, the demand for potable water in Singapore increases steadily, the former way of water supply can't satisfy the need of people's living and production. To augment water supply, Singapore has built the Marina Barrage to develop further its rainfall (see Figure 7).

Figure 7. Marina Barrage in Singapore (Across Marina Channel at Marina South)

The Marina Barrage is a dam built across the 350-metre wide Marina Channel to keep out seawater. Marina Reservoir, together with Punggol and Serangoon reservoirs, increased Singapore's water catchment area from half to two-thirds of Singapore's land area in 2011. It creates a fresh water reservoir behind it at the same time acts as a tidal barrier that prevents high tides from causing flooding of inland low-lying areas. The Project is unique in that it is designed to achieve three aims: to act as a tidal barrier for flood control, to create a new reservoir to augment the water supply and to maintain a new body of freshwater at constant level in the heart of the city as a major lifestyle attraction. The Project has been carefully designed to blend in well with the environment, with guidance from Urban Redevelopment Authority (URA)'s Design Advisory Panel. Figure 8 shows the model of the Marina Barrage design, and Figure 9 shows the working principal of the Marina Barrage.

The barrage, which comprises nine numbers of 26.8-metre-long hydraulically operated steel crest gates, will be built across the 350m wide Marina Channel to keep out sea water. Under normal conditions, the steel gates will remain closed to isolate the reservoir from the sea. During heavy rain, the steel gates will open as necessary to release excess stormwater to the sea when the tide is low. However, when it is not possible to do so during high tide, the Drainage Pumping Station capable of pumping up to 280 cubic metres per second will pump out the excess stormwater into the sea.

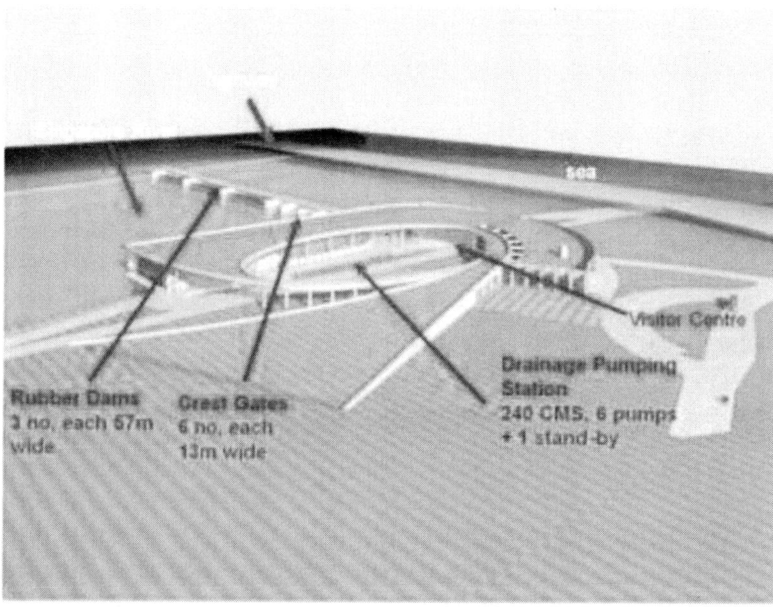

Figure 8. Model of the Marina Barrage Design

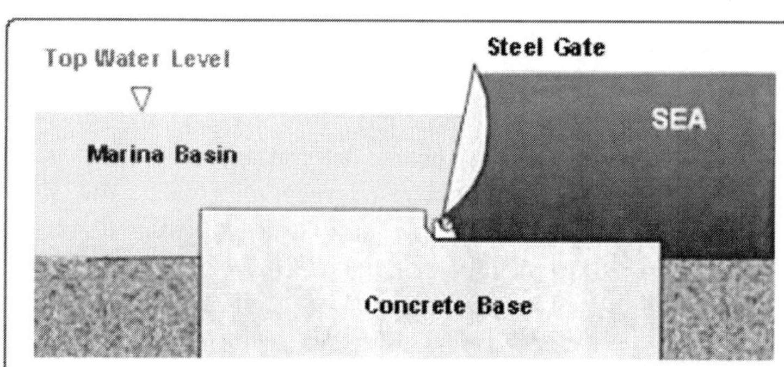

Figure 9. Working Principal of the Marina Barrage

For water supply, the Marina Barrage have enhanced Singapore's water supply in line with Singapore's Four National Taps water supply strategy to diversify its water sources (The 4 National Taps are: local catchment, reclaimed water, desalted water and imported water). The Marina Reservoir will have the largest urban catchment of 10,000 ha among all the reservoirs. With this Project, about 60 per cent of Singapore will become catchment area.

For flood control, the Marina Barrage is part of a comprehensive flood control scheme to alleviate existing flooding in the low-lying areas of the city such as Chinatown, Boat Quay, Jalan Besar and Geylang. During heavy rain, the series of nine crest gates at the dam will be activated to release excess storm water into the sea when the tide is low. In the case of high tide, giant pumps which are capable of pumping an Olympics-size swimming pool per minute will drain excess storm water into the sea. With the Barrage and other flood-alleviation projects, flood-prone areas in Singapore will be further reduced from the current 150 ha to 85 ha, down from 3200 ha in the 1970s.Figure 9 shows the principal of how the barrage control the flood.

For lifestyle attraction, it is ideal for all kinds of recreational activities such as boating, windsurfing, kayaking and dragon boating etc. As the water in the Marina Basin is unaffected by the tides, its water level will be kept constant all year round.Marina Barrage is a showpiece of environmental sustainability, and won the Green Mark Platinum Infrastructure Award, the top award at the BCA Awards organised by the Building and Construction Authority in May 2009.

4. ENVIRONMENTAL AND SOCIAL IMPACT

The coastal reservoir at river mouth can capture every single drop of runoff, and also has the potential to collect all contaminants yielded from the catchment. For water quality, coastal reservoirs catch the stormwater runoff which is closer to drinking water in quality than the treated salt water is that is produced by desalination. Only high quality water that is free of contaminates will be allowed to enter the coastal reservoir for later use, all poor quality water will drained into the ocean guaranteeing an acceptable level of water quality.

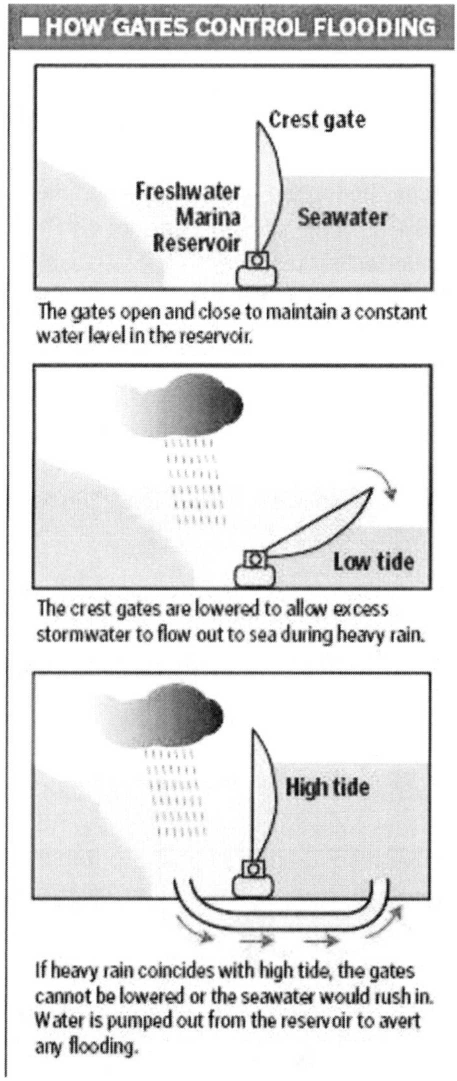

Figure 10. Principal of How the Barrage controls Flood

For environmental impact, as there is nearly no land requirements for building coastal reservoirs and water quality can be guaranteed, the damage to the local ecosystems and marine wildlife can be minimized greatly. The proposed method of harvesting will avoid any severe environmental impacts such as depriving local rivers of water. As the reservoirs construction at sea, no land is needed for construction purposes making them better for the environment then typical inland mountainous reservoirs. Only high quality water (free of contaminates) will be allowed to enter the reservoir, which will help to minimize the impact of the coastal reservoir on the local ecosystem and marine wildlife.

Due to the reservoirs ability to capture every single drop of runoff it will always be functional as long as rain continues to fall. This allows for a stormwater capturing technique that is highly sustainable, will last for many decades and does not pose any major risks to the population. Coastal reservoirs will not lead to flooding, as the water collected is in the ocean. When there is too much stormwater which may occur during long wet periods, the excess water will be drained into the ocean this preventing any form of flooding. For place making, coastal reservoirs are able to fit into any location at the mouth of a river (river estuary) entering into ocean. This is very convenient for coastal cities which have huge population and less space.

5. COMPARISON BETWEEN COASTAL RESERVOIR AND OTHER WATER SOLUTIONS

Energy use: Energy use on coastal reservoir is mainly on the transfers of potable water to areas above sea level from a water treatment plant like desalination plantthat also needs to transfer potable water to users above sea level.Water Recycling from a wastewater treatment plant needs more energy as it needs to be transported to inland reservoirs for mixing before it is distributed to users, this incurs high energy cost of mass water transport.

Emigrant cost: as coastal reservoirs are usually built near to the estuary in the sea, there is almost no emigrant cost. But inland reservoirs can affect large amounts of people especially when damming major valleys. For desalination, water recycling and mass water transport, there is no need to emigrate too many people while some infrastructure may still require property acquisition.

Waste output: coastal reservoirs and inland reservoirs have no waste output with comparison of desalination, which has high wastage from desalination (as little as 30% freshwater yields from seawater intake). Waste output of water recycling is mainly from water treatment waste.

Usable lifespan: coastal reservoirs are affected by siltation and rainfall variability, the same as inland reservoirs. As seawater will always be available, desalination mainly depends on sustainable energy supply. Water recycling which is similar to desalination, as waste water will always be available, depends on sustainable energy supply, which is similar to desalination. For mass water transport, it depends on both reliability of source and sustainable energy supply.

Maintenance cost: The maintenance cost on coastal reservoirs and inland reservoirs is low as it in mainly used for general dam maintenance, besides extra cost on coastal reservoirs includes coastal erosion / salt affects' protection. For desalination, exceptcoastal erosion / salt affects protection, the maintenance also includes treatment systems, servicing and parts replacement. So do the maintenance of water recycling. High maintenance of mass water transport cost on transport device, water transfer systems and management plans / agreements.

Potential for further capability change: coastal reservoirs' potential is high for the storage volume can be expanded at most sites, while inland reservoirs can't be changed very muchas initial design is usually for maximum possible yield. It is also not difficult for desalination to enlarge or to reduce water yield, which only needmore desalination plants can be built at the coast if energy is available. Waste water recycling is similar to desalination, which need more waste water treatment built. But potable reuse is still a distant possibility and may never be implemented except under extreme conditions [32]. For mass water transport, it depends on availability of remote sources and availability of energy.

Table 3 shows the comparison of construction cost and cost per kilolitre of water between coastal reservoir and other water solutions

Table 3. Relative comparison of Coastal Reservoirs to other Australia Water Resources

	Coastal Reservoirs	Inland Reservoirs	Desalination	Water Recycling	Mass water transport
Construction cost per kilolitre of water(US$)	2.67-6.01	5.83-7.5	6.41-10.08	5.57-8.30	2.75-6.37
Cost per kilolitre of water(US$)	0.15-0.25	0.34-0.4	0.43-1.13	1.44-1.53	0.39-6.1

The data of the table is based on the conditions in Australia. The construction cost per kilolitre of waterand cost per kilolitre of water of coastal reservoirs are calculated based on the existing coastal reservoirs in the world [33-37]. For inland reservoir, the data is based on statistical data of the UnitedNations [38]. For desalination, the construction cost per kilolitre of waterand cost per kilolitre is from comprehensive evaluation of Israel [39], Singapore [40], and Australia [41],America [42-43], India [44-45] and so on, which in the world mainly have robust desalination technology. For water recycling, the data is on account ofRemco Engineering on Water Systems and Controls in USA [46] and National Snapshot of Current andPlanned Water Recycling and ReuseRates in Australia [47]. The data resource about mass water transport is in view of water transport project in Australia [48], the USA [49-50], China [51] and Africa, which of the countries in the world are famous for water transporting. From the above, coastal reservoirs are low-cost compared with other water solutions.

6. Conclusions

"Water is everybody's business".

One of the Key Messages in the 2nd World Water Forum

Different from the proposals of desalination and wastewater recycling, the proposal of coastal reservoir will use the natural catchment runoff and there will be no need to separate the freshwater and salt or wastes, thus the energy cost associated with the treatment will be zero, there will be no carbon dioxide emission either. The pumping costs for proposals of desalination, wastewater recycling and coastal reservoir will be almost same and negligible relative to their treatment cost, thus no comparison will be given in this study. When compared with the proposal of on-land reservoirs, the method of coastal reservoirs has no cost to cover the inundation of land and people's relocation, normally this is very expensive and could be more than half of dam's construction cost. The existing coastal reservoirs in Hong Kong that have existed over 50 years, have no significant impacts on the ecosystem, and also no evidence from other coastal reservoirs in China, Korea and Singapore etc, shows that coastal reservoirs have significant environmental impacts. This conclusion could be valid for the proposed coastal reservoir in this study as fish will still have the passage to go to upstream; every year only the excessive floodwater will be diverted and the time and amount of water diversion could be adjusted to reach the win-win solution for water resources development and ecosystem like fish breeding and stocks alike. It should be stressed that experts from different disciplines should be invited to clarify the environmental impacts of coastal reservoir.

The main advantage of coastal reservoir is storing excess fresh water in the rainy season, and then the fresh water can be transferred to near watersheds through artificial channels or pipelines. Last, the fresh water can be supplied people to drink and meet the needs of agriculture [31].These do not require mountainous areas for dam construction. The highest growth in population and therefore water demand is usually expected to be close to the coast. Coastal reservoir can be located close to the high demand area.

Water demand is related to population growth, in theory population increases in coastal areas are much higher relative to the inland regions, and hence the water demand in coastal regions is likely to continue to be higher than inland regions. The future water demand in coastal regions is difficult to meet by developing more on-land reservoirs as it needs an ideal combination of suitable hydrological, geological and topographic conditions. Therefore, coastal reservoir will play a more and more important role in the freshwater resources development in the near future.

ACKNOWLEDGEMENTS

The research presented in this paper has been supported by the University of Wollongong, Australia. The works are also supported, in part, by the open fund (SKLH-OF-1002) provided by State Key Laboratory of Hydraulics and Mountain River Engineering at Sichuan University, and the National Natural Science Foundation of China (51061130547) and (51179015), (51228901) .

REFERENCES

1. WorldWaterCouncil. Water Crisis. http://www.worldwatercouncil.org/index.php?id=25 (accessed 22 July 2012).

2. WHO/UNICEF. Water Supply and Sanitation. 2012.

3. WaterAmbassadors. Water Facts. http://www.waterambassadors canada.org/water_crisis.html (accessed 22 July 2012).

4. Lubin G. MAP OF THE DAY: The World Water Crisis. http:// www.businessinsider.com/water-crisis-2011-3#ixzz1wGBzeUdf (accessed 22 July 2012).

5. Arnell NW. Climate Change and Global Water Resources: SRES Emissions and Socio-economic Scenarios. Global Environmental Change 2004; 14(1) 31-52.

6. Shah T, Singh O and Mukherji A. Some aspects of South Asia's Groundwater Irrigation Economy:Analyses from ASurvey in India, Pakistan, Nepal Terai and Bangladesh. Hydrogeology Journal 2006; 14(3) 286-309.

7. Fedoroff N, Battisti D, Beachy R, Cooper P, Fischhoff D, Hodges C, Knauf V, Lobell D, Mazur B and Molden D. Radically Rethinking Agriculture for the 21st Century. Science 2010; 327(5967) 833-834.

8. Hanjra MA and Qureshi ME. Global Water Crisis and Future Food Security in an Era of Climate Change. Food Policy 2010; 35(5) 365-377.

9. CSIRO. Water Availability in the Murray–Darling Basin: A Report from CSIRO to the Australian Government. http://www.csiro.au/files/ files/ po0n.pdf (accessed 22 July 2012).

10. UNDP. Human Development Report 2011.

11. BluePlanetNetwork. The Keystone to Transform a Life. http:// blueplanetnetwork.org/water/firststep (accessed 22 July 2012).

12. Cabral KP, Limcuando VRL and Bontia JS. A Preliminary Study on the Capability of Burned Rice Hulls to Reduce Escherichia Coli in Drinking Water. 2007.

13. Liu J, Temporal and Spatial Distribution of Population and Disaster Exposure Assessment in LECZ, China, in Environmental Geography. 2010, Shanghai Normal University: China. 41-42.

14. NOAA. The U.S. Population Living in Coastal Watershed Counties. http://stateofthecoast.noaa.gov/population/welcome.html (accessed July 26, 2012).

15. Hatton T. Population Growth and Urban Development. 2011.

16. NationalSeaChangeTaskforce. A 10-point plan for coastal Australia: towards a sustainable future for our coast. 2010.

17. Creel L, Ripple Effects: Population and Coastal Regions. 2003, Population Reference Bureau

18. Chanson H. JDP. Railway Dams in Australia: Six Historical Structures. Transactions Newcomen Society 1999; 71(b) 283-304.

19. Sadhwani JJ, Veza JM and Santana C. Case studies on environmental impact of seawater desalination. Desalination 2005; 185(1–3) 1-8.

20. McCutcheon JR, McGinnis RL and Elimelech M. A Novel Ammonia—Carbon Dioxide Forward (direct) Osmosis Desalination Process. Desalination 2005; 174(1) 1-11.

21. Van der Bruggen B and Vandecasteele C. Distillation vs. Membrane Filtration: Overview of Process Evolutions in Seawater Desalination. Desalination 2002; 143(3) 207-218.

22. Lattemann S,Höpner, Thomas. Environmental Impact and Impact Assessment of Seawater Desalination. Desalination 2008; 220(1–3) 1-15.

23. WHO. Desalination for Safe Water Supply, Guidance for the Health and Environmental Aspects Applicable to Desalination.

24. S. Lattemann TH. Seawater Desalination—Impacts of Brine and Chemical Discharges on the Marine Environment. Desalination Publications, L'Aquila, Italy 2003.

25. MEDRC. Assessment of the Composition of Desalination Plant Disposal Brines. 2002.

26. Lattemann S. Protecting the Marine Environment. Seawater Desalination 2009; 273-299.

27. UNEP. Seawater Desalination in Mediterranean Countries: Assessment of Environmental Impact and Proposed Guidelines for the Management of Brine. 2001.

28. Bdour A.N. H, M.R., Tarawneh, Z. Perspectives on sustainable wastewater treatment technologies and reuse options in the urban areas of the Mediterranean region. Desalination 2009; 237(1-3) 162-174.

29. Goonrey CM, Lechte P, Maheepala S, Mitchell VG and Perera BJC. Examining the Technical Feasibility of Using Stormwater as an Alternative Supply Source within an Existing Urban Area--ACase Study. Australian Journal of Water Resources 2007; 11(1) 13-29.

30. Shu-Qing Yang SF. Coastal Reservoirs Can Harness Stormwater. Water Engineering Australia 2010 25-27.

31. Yang S, Coastal reservoir construction. 2004.

32. Asano T. Water From(waste) water- the dependable water resource. Water Science & Technology 2002; 45(8) 24.

33. Liu X, Wang X, Guan X, Kong L and Su A. Study on Storage Capacity and Characteristic Water Level for Qingcaosha Reservoir [J]. Water Resources and Hydropower Engineering 2009; 7 002.

34. Cho Do. The Evolution and Resolution of Conflicts on Saemangeum Reclamation Project. Ocean & Coastal Management 2007; 50(11) 930-944.

35. Lie Hj, Cho Ch, Lee S, Kim Es, Koo BJ and Noh Jh. Changes in Marine Environment by a Large Coastal Development of the Saemangeum Reclamation Project in Korea. Ocean and Polar Research 2008; 30(4) 475-484.

36. Lee SI, Kim BC and Oh HJ. Evaluation of Lake Modification Alternatives for Lake Sihwa, Korea. Environmental Management 2002; 29(1) 57-66.

37. Onn LP. Water Management Issues in Singapore. Water in Mainland Southeast Asia. 2005; 29.

38. UN, UNdata. 2012.

39. Sitbon S, French-Run Water Plant Launched in Israel in European Jewish Press. 2005.

40. Veatch B, Designed Desalination Plant Wins Global Water Distinction, in Black & Veatch. 2006.

41. WaterTechnology. Perth Seawater Desalination Plant, Australia http://www.water-technology.net/projects/perth/ (accessed September 28th 2012).

42. Sweet P, Desalination Gets a Serious Look, in Las Vegas Sun. 2008.

43. Kranhold K, Water, Water, Everywhere, in the Wall Street Journal. 2008.

44. SistlaPVSea, Low Temperature Thermal Desalination Plants, in Proceedings ofthe Eighth ISOPE Ocean Mining Symposium, Chennai, India. 2009: India.

45. Chennai, Low Temperature Thermal Desalination Plants Mooted in the Hindu. 2007.

46. Remco, Water Recycling Costs 2012, Remco Engineering on Water Systems and Controls.

47. Kym Whiteoak RB, NadjaWiedemann. National Snapshotof Current and Planned Water Recycling and Reuse Rates. 2008.

48. HWA. Piping Water From the Ord to Perth. 2003.

49. Oksche A, Schülein R, Rutz C, Liebenhoff U, Dickson J, Müller H, Birnbaumer M and Rosenthal W. Vasopressin V2 Receptor Mutants that Cause X-linked NephrogenicDiabetes Insipid us: Analysis of Expression, Processing, and Function. Molecular Pharmacology 1996; 50(4) 820-828.

50. Lavack C. Hydrological Changes in the Sierra Nevadas due to Loss in Glacierized Area. 2010.

51. Berkoff J. China: The South-North Water Transfer Project--Is It Justified? Water Policy 2003; 5(1) 1-28.

Chapter 9

Application of SWAT Model to the Olifants Basin: Calibration, Validation and Uncertainty Analysis

Charles Gyamfi[1,2*], Julius Musyoka Ndambuki[1], Ramadhan Wanjala Salim[1]

[1]Department of Civil Engineering, Tshwane University of Technology, Pretoria, South Africa

[2]School of Bio-Resources Engineering, Anglican University College of Technology, Nkoranza, Ghana

ABSTRACT

The application of the Soil and Water Assessment Tool (SWAT) to the Olifants Basin in South Africa was the focus of our study with emphasis on calibration, validation and uncertainty analysis. The Basin was discretized into 23 sub-basins and 226 Hydrologic Response Units (HRUs) using 3 arc second (90 m × 90 m) pixel resolution SRTM DEM with stream gauge B7H015 as the Basin outlet. Observed stream flow data at B7H015 were used for model calibration (1988-2001) and validation (2002-2013) using the split sample approach. Relative global sensitivity analysis using SUFI-2 algorithm was used to determine sensitive parameters to stream flow for calibration of the model. Performance efficiency of the Olifants SWAT model was assessed using Nash-Sutcliffe (NSE), coefficient of determination (R^2), Percent Bias (PBIAS) and Root Mean Square Error-Observation Standard deviation Ratio (RSR). Sensitivity analysis revealed in decreasing order of significance, runoff curve number (CN2), alpha bank factor (ALPHA_BNK), soil evaporation compensation factor (ESCO), soil available water capacity (SOIL_AWC, mm H_2O/mm soil), groundwater delay (GW_DELAY, days) and groundwater "revap" coefficient (GW_REVAP) to be the most sensitive parameters to stream flow. Analysis of the model during the calibration period gave the following statistics; NSE = 0.88; R^2 = 0.89; PBIAS = −11.49%; RSR = 0.34. On the other hand, statistics during the validation period were NSE = 0.67; R^2 = 0.79; PBIAS = −20.69%; RSR = 0.57. The observed statistics indicate the applicability of the SWAT model in simulating the hydrology of the Olifants Basin and therefore can be used as a Decision Support Tool (DST) by water managers and other relevant decisions making bodies to influence policy directions on the management of watershed processes especially water resources.

KEYWORDS

Calibration, Validation, Uncertainty Analysis, Olifants Basin, SWAT Model

1. INTRODUCTION

Like many other basins in the world [1] - [4] , the Olifants Basin in South Africa is faced with challenging watershed issues ranging from water quantity, quality and the influence of anthropogenic activities on available water resources [5] . These watershed problems have over the years increasingly gained global attention owing to the fact that water resources are professed to be the pivot for which developmental agenda and improvement in livelihoods can be achieved [6] . Others [7] [8] have suggested that for improvements in livelihood in terms of poverty reduction, livelihood empowerment and long term economic growth to be attained, then there would be the need to have access to freshwater that meets both quantity and quality demands. More importantly, also, is the fact that issues of water are now perceived to be of national and regional security risks [9] . As water resources diminish due to population growth, pollution and over exploitation, the competition for water will continually exacerbate which is seen as a potential source of conflict [10] - [12] . In line with meeting the water supply needs of man and the ecosystem, a new paradigm shift for water resources planning has therefore been suggested [13] [14] .

To effectively plan for water resources and mitigate against existing watershed problems calls for a clear understanding and insight into prevailing watershed processes in order to spearhead planning and implementation of appropriate control measures. One of such ways has been the application of tools such as hydrologic models, water quality models and geographic information systems which have proven versatile in the acquisition of knowledge in line with watershed processes [15] . However, the use of tools such as hydrologic models presents some difficulties due to the large number of parameters involved in the hydrologic modeling process coupled with measurement errors in model inputs [16] [17] . For a hydrologic model to be reliable in the planning processes it is required that resulting uncertainty in its application be curtailed within the barest allowable range through a reliable calibration and validation procedure. In view of this, we present in this work an outlook on the application of the Soil and Water Assessment Tool (SWAT) a semi distributed hydrologic model to simulate the stream flow of the Olifants Basin with emphasis on 1) identification of sensitive parameters to stream flow; 2) calibration and validation of the model for stream flow for the study basin; and 3) an assessment of model performance. The objects of the study were to test the applicability of the SWAT model to the Olifants Basin and to examine its potential as a decision support tool for water managers and policy makers.

The Soil and Water Assessment Tool (SWAT) is a physically based medium to large scale river basin model developed by the Agricultural Research Service of the United States Department of Agriculture (USDA-ARS) [18] . The basic operational unit of the SWAT model is the hydrologic response units (HRUs) which consist of an area of homogenous land use, topography and soil characteristics. The concept of HRUs is to cater for the diverse ways by which different land use and soil types impacts on hydrology of a watershed. Since its introduction, SWAT has been used in diverse applications ranging from water resources planning, investigation into hydrological processes, water quality modeling and climate change studies [19] - [22] . For details on the SWAT model one is referred to Neitsch et al. [23].

2. MATERIALS AND METHODS

2.1. Study Area

The Olifants Basin was used as the case study due to the socio-economic recognition of this basin to the South African economy but however reported to be water stressed calling for an effective action of water resources planning and management. The Olifants basin which is divided into two by an escarpment is located in the north-eastern part of South Africa (Figure 1). Geographically, it is bordered by latitudes 23.8˚S and 26.5˚S and longitudes 28.3˚E and 31.9˚E. It traverses three provinces in South Africa namely, Gauteng, Mpumalanga and Limpopo where it finally enters into Mozambique. The Olifant River with a main stem of approximately 770 km in length drains a total land surface area of 54,308 km^2 [24] . The River originates from Trichardt to the east of Johannesburg in the province of Gauteng and then flows in the north-eastern direction through the provinces of Mpumalanga and Limpopo. The Olifant River crosses the Mozambique border and flows into the Massingire dam. The main tributaries of the Olifant are the Wilge, Moses, Elands and Ga-Selati rivers on the left bank and Klein Olifant, Steelpoort and Blyde rivers on the right bank [25] . On the right bank, steelpoort is the largest tributary draining an area of 7136 km^2 compared to the Elands on the left bank with a drainage area of 6148 km^2. Generally, the elevation of the basin ranges from 0 - 2400 meters above mean sea level (masl). Climatically, the basin is characterized by the movement of air masses that is known to be associated with Inter-Tropical Convergence Zone (ITCZ). In summer, temperatures range from a minimum of 18˚C to a maximum of 34˚C whereas in winter temperatures ranges from 5˚C to 26˚C [25] . Rainfall is seasonal in the basin, occurring during the months of October to April with appreciable spatio-temporal variability. The mean annual precipitation (MAP) is documented by Schulze et al. [26] to be 630 mm with peaks in January.

2.2. Scope of the Study

The scope of the study as inferred in this work refers to the section of the Olifants Basin upstream of the stream gauge station B7H015. This section of the basin drains a total area of 50,024 km^2. The remaining section of 4, 200 km^2 downstream of gauge B7H015 was not included in this study. As the focus of the study was on calibrating and validating the SWAT model, it was imperative therefore to select a reliable stream flow gauge station for calibration and validation purposes hence the selection of gauge B7H015 with considerable data from 1980-2013.

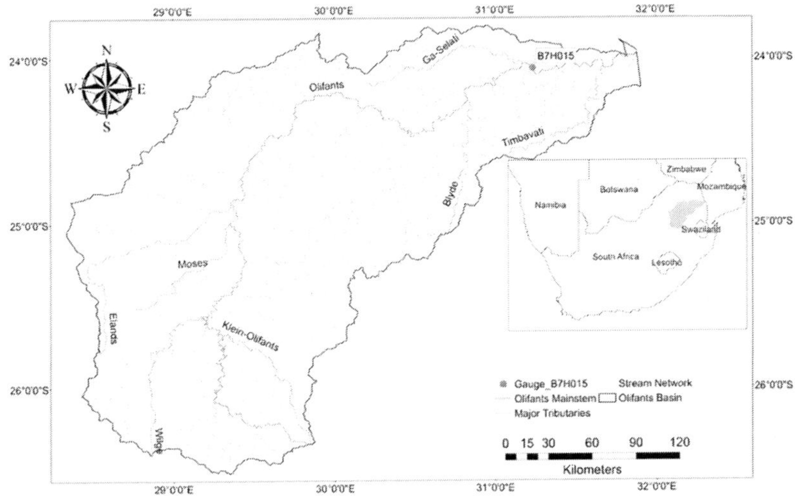

Figure 1. Location of the Olifants Basin.

2.3. Data Sources

2.3.1. Hydro-Meteorological Data

From the Department of Water Affairs (DWA), South African Weather Service (SAWS) and the Climate Forecast System Reanalysis (CFSR) database, climatic variables inclusive of maximum and minimum temperature, rainfall, solar radiation, wind speed and humidity were acquired. These variables served as the atmospheric parameters used in driving the hydrologic simulation of the study watershed. Meteorological stations were selected if they satisfied the criterion of not having more than 10% missing data. Based on the set criteria and taking into cognizance the time span, 13 meteorological stations were finally selected for use (Figure 2). With the exception of meteorological station 04774946, all the other stations had varying degree of missing data which were patched prior to using in the hydrologic simulations. Stream flow gauge readings were also acquired from DWA for model calibration and validation.

2.3.2. Soil Data

Soil physico-chemical and hydrological properties were obtained from various FAO-UNESCO soil databases [27] [28] . The soil databases contains attributes for each soil polygon, namely, soil texture, Hydrological Soil Group (HSG), soil depth, rock fragments and organic carbon content and these databases are considered the single most comprehensive available resource with vast database on soil attributes for Africa and the world [29] .

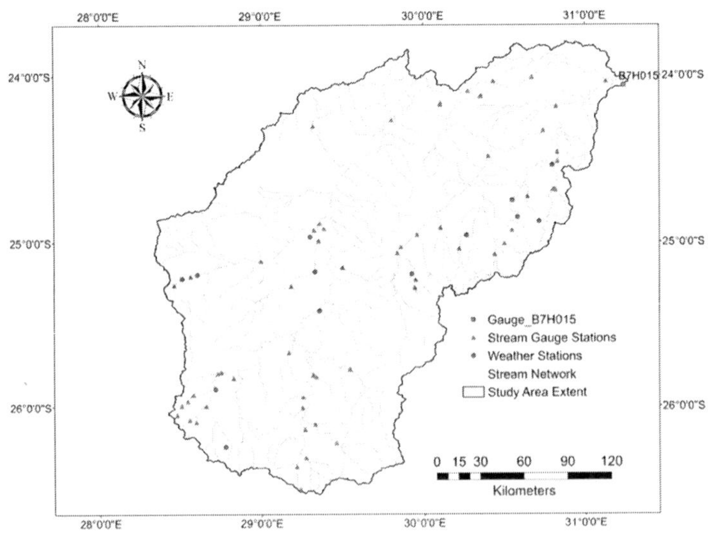

Figure 2. Hydro-meteorological stations within the study area.

A subset of the soil map for the study area showed five major soil types. Presented in Table 1, are the major soil types with some selected physical and hydrological properties that were used in the study. All the attributes of the soils (soil texture, Hydrological Soil Group (HSG), soil depth, rock fragments and organic carbon content) were captured into a database using GIS which provided a link for usage in SWAT. The value fields that ensured the proper recall of gridded soil information were the serial number (SNUM) and the FAO soil code.

2.3.3. Land Use/Cover Data

Landsat 7 ETM/ETM + level 1 terrain corrected satellite images obtained from the Global Visualization Viewer (GloVIS) formed the basis for the execution of the land use and land cover classification component of the study. The images which are of spatial resolution 30 m, were obtained for Path/Row; 168/077, 169/077, 169/078, 170/077 and 170/078 for the period 2013. The images were classified into five-level 1 classes based on the land cover and land use classification system developed by Anderson et al. [30] for the interpretation of remote sensor data at various scales and resolutions.

2.4. Model Setup and Parameterization

Delineation of the study watershed (Figure 3) was done in Arc SWAT using the automatic watershed delineator with a 3 arc second (90 m × 90 m) pixel resolution SRTM DEM. The SRTM DEM version 4 which was released by the Consortium for Spatial Information of the Consultative Group of International Agricultural Research (CGIAR-CSI) was selected for use due to its significant improvements over previous versions. Watershed delineation was guided by the procedures used by Neitsch et al. [31] . Stream gauge B7H015 was selected to be the outlet point of the watershed. This gauge station is located on the main-stem of the Olifants River. The selection of this gauge station was informed by the availability of stream flow data that exist for this gauge and as result deemed to be appropriate for model calibration and validation. The area of the delineated watershed was cut short of previously defined Olifants Basin due to the gauge station that was used in defining the watershed outlet.

The watershed was discretized into a total of 23 sub-basins with the smallest draining an area of 45 km^2 and the largest an area of 6681 km^2. Total drainage area of the delineated watershed was 50,024 km^2 with the main stem of the Olifants traversing a distance of 626 km. A comparison of the delineated Olifants Basin used in this study to that already existing revealed a drainage area difference of approximately 4200 km^2. This section of the Olifants located downstream of the delineated watershed was not considered in this study.

Table 1. Physical and hydrological properties of major soils in the Olifants Basin.

Soil type	FAO code	Hydrologic group	Texture	BD (g/cm^3)*	AWC (mm/mm)*	Organic carbon content (%)
Cambic Arenosols	Qc42-1a	C	Sandy loam	1.45	0.08	0.35
Chromic Luvisols	Lc65-1-2a	C	Sandy loam	1.45	0.14	0.55
	Lc3-2ab	C	Sandy Clay Loam	1.50	0.18	0.50
	Lc66-1a	C	Sandy Loam	1.45	0.14	0.40
Chromic Vertisols	Vc23-3a	D	Clay	1.60	0.14	0.50
	Vc1-3a	D	Clay	1.65	0.13	0.50
Orthic Acrisols	Ao69-1a	C	Sandy Loam	1.50	0.01	0.50
Rhodic Ferralsols	Fr20-3bc	C	Clay	1.20	0.16	1.10

BD, bulk density; AWC, available water content, * Average for layer 1 and 2 of soil profile.

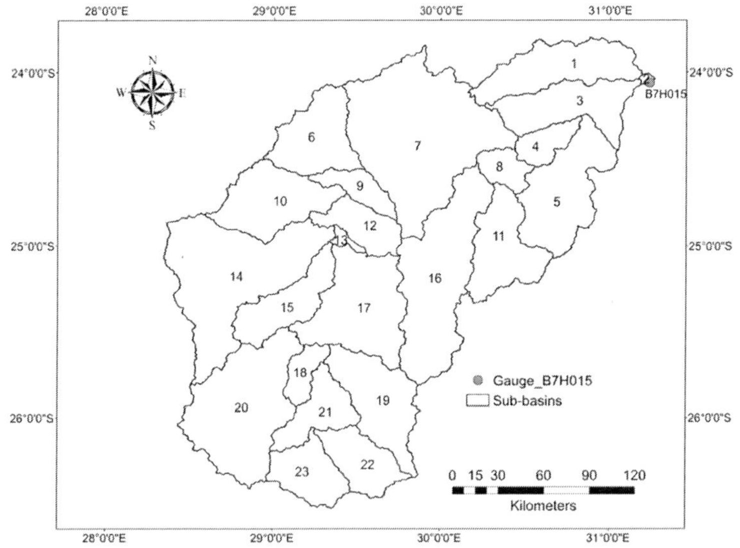

Figure 3. Discretization layout of the Olifants Basin using gauge B7H015.

2.4.1. Slopes and HRU Definition

Slopes discretization of the Olifants watershed was guided by the classification used by FAO [32] . Based on this, the study watershed was classified to have three slope classes namely; level to gently undulating (<8%), rolling to hilly (8% - 30%) and steeply dissected to mountainous (>30%). The multiple slope classes approach was used to cater for the different slopes that exist within the area unlike using a single slope approach.

In defining the HRU, an overlay analysis of the land use, soils and slopes were carried out and a unique combination of a particular type of land use, soils and slopes were then grouped to form an HRU. The HRU relying on the distinct characteristics of land use, soils and slopes of the watershed forms the premise for the prediction of watershed processes. To account for complex watershed processes without undue attention for minor physiographic features, multiple HRUs with threshold definitions of 10%, 10%, 10% for land use, soils and slopes were used respectively. The implication of the threshold value is to eliminate in the generation of the HRUs all areas with coverage less than the stipulated HRU threshold. In doing so, the model simulates for watershed processes based on the substantive land use, soils and slopes within in the watershed.

2.4.2. Parameter Sensitivity Analysis

As watershed processes are influenced by a large number of parameters, sensitivity analysis was performed using the Sequential Uncertainty Fitting (SUFI-2) algorithm to identify the key parameters that affect stream flow for calibration. The global sensitivity analysis approach which considers the sensitivity of one parameter in relation to other parameters under consideration was used to determine the sensitive parameters in this study [33] . The t-stat which gives a measure of the sensitivity of a parameter and the p-value, the significance of the sensitivity of that parameter were used to rank the various parameters considered to influence stream flow. From other studies the listed parameters in Table 2 were discovered to be sensitive to stream flow [34] [35] . However, these parameters vary from one watershed to the other depending on prevailing geomorphologic characteristics and activities existing in a watershed and therefore cannot assume generalization status. In initializing the sensitivity analysis, the parameters listed in Table 2 were considered and the final parameters selected based on the t-stat and the p-value. Following sensitivity analysis, model calibration was done using the selected most sensitive parameters.

2.4.3. Model Calibration, Validation and Performance Evaluation

The complex processes occurring in watersheds coupled with the uncertainty inherent in hydrologic modeling parameters, inputs and measured data requires that hydrologic models be calibrated and validated to minimize predictive errors [35] . One of the most widely accepted algorithm; the Sequential Uncertainty Fitting (SUFI-2) which operates based on the Latin Hypercube sampling procedure was used to calibrate and validate the hydrologic setup [35] [36] . A split sample procedure using monthly stream flow data from stream gauge B7H015 for the period 1988-2001 and 2002-2013 were used for calibration and validation respectively. The first 8 years were used as warm up period to mitigate unknown initial conditions. As a result, the first 8 years were excluded from the analysis.

Table 2. Parameters considered in sensitivity analysis.

Parameter name	Description
CN2	SCS runoff curve number
ESCO	Soil evaporation compensation factor
GWQMN	Threshold depth of water in the shallow aquifer required for return flow to occur (mm H_2O)
SOL_AWC	Soil available water storage capacity (mm H_2O/mm soil)
GW_REVAP	Groundwater revap coefficient
RCHRG_DP	Deep aquifer percolation fraction
SOL_Z	Soil depth (mm)
SURLAG	Surface runoff lag coefficient (days)
SOL_K	Soil conductivity (mm/h)
CH_K2	Effective hydraulic conductivity in the main channel (mm/h)
ALPHA_BF	Baseflow alpha factor (days)
GW_DELAY	Groundwater delay (days)
ALPHA_BNK	Baseflow alpha factor for bank storage (days)
REVAPMN	Threshold depth of water in the shallow aquifer for "revap" to occur (mm)

The SUFI-2 algorithm was operationalized within the SWAT calibration and Uncertainty Procedure (SWAT- CUP) environment. As suggested by Abbaspour et al. [36] several model simulations were executed with a minimum of 500 simulations in each run. Following each run, simulated results were compared with observed variables of interest and the performance of the model in simulating the observed variable judged against four objective functions. The choice of an objective function is dictated by the object of any particular study and hence there is no objective function that is universally applicable to all situations [37] . In this study however, efficiency measures including the Nash-Sutcliffe (NSE) [38] [Equation (1)], the coefficient of determination (R^2) [8] [Equation (2)], percent bias (PBIAS) [39] [Equation (3)] and RMSE-observations standard deviation ratio (RSR) [39] [Equation (4)] were used to evaluate the performance of the model.

$$NSE = 1 - \frac{\sum_{i=1}^{n}(O_i - S_i)^2}{\sum_{i=1}^{n}(O_i - \bar{O})^2} \quad (1)$$

$$R^2 = \left[\frac{\sum_{i=1}^{n}(O_i - S_i)(S_i - \bar{S})}{\left(\sum_{i=1}^{n}(O_i - \bar{O})^2\right)^{0.5}\left(\sum_{i=1}^{n}(S_i - \bar{S})^2\right)^{0.5}} \right]^2 \quad (2)$$

$$PBIAS = \frac{\sum_{i=1}^{n}(O_i - S_i)\times 100}{\sum_{i=1}^{n}O_i} \quad (3)$$

$$RSR = \frac{\sqrt{\sum_{i=1}^{n}(O_i - S_i)^2}}{\sum_{i=1}^{n}(O_i - \bar{O})^2} \quad (4)$$

where;

O_i = observed variable, S_i = simulated variable, $\bar{O}$= mean of observed variable, $\bar{S}$= mean of simulated variable, n = number of observations under consideration.

3. RESULTS AND DISCUSSION

3.1. Characteristics of Discretized Watershed

Discretization of the Olifants watershed resulted in a total of 23 sub basins with 226 HRUs using land use, soil and slope thresholds of 10%, 10% and 10% respectively. Elevation ranges between 0 - 2328 m with a mean of 1220 m and a standard deviation of 378 m. Characteristics of the delineated watershed with respect to land use type, soils and slope is summarized in Figure 4 andTable 3. The land use characterization for 2013 based on the stated threshold values consisted mainly of range land (40.05%), agriculture (36.88%), residential (22.62%) and mixed forest (0.46%). Agriculture being the second largest land use type is in affirmation of the economic importance this area contributes towards the gross domestic product (GDP) of the South African economy. The distribution of soil types recorded chromic luvisols to be the most dominant soil with a total coverage of 38.81% of the watershed followed by cambic arenosols (33.03%) and chromic vertisols (21.21%). Orthic acrisols and rhodic ferralsols constituted a marginal proportion of 5.77% and 1.18% respectively. Slope classification following FAO slope classes indicated a larger proportion (76.57%) of the studied watershed having a level to gently undulating terrain within slope class <8%. Rolling to hilly terrain (slope class 8% - 30%) constituted 18.26% with the remaining 5.17% of the area being steeply dissected to mountainous (slope class >30%). Inference from the classified slopes generally portrays that the Olifants watershed has varying terrain with approximately 24% of the area having slopes greater than 8%. The steep slopes (>30%) are located in the region of the escarpment.

3.2. Initial Model Run Analysis

An initial model simulation was executed for the calibration period 1988-2001 and the result presented in Figure 5. The time series analysis of the observed flow (Figure 5(a)) suggests significant seasonal flow during the wet months of January to April and October to December. The trend observed is expected because during this time periods the Olifants Basin experiences rainfall which translates into the increase in stream flow. The model simulations captured the trend in the observed stream flow; however, it overestimated the observed stream flow time series by over 70%.

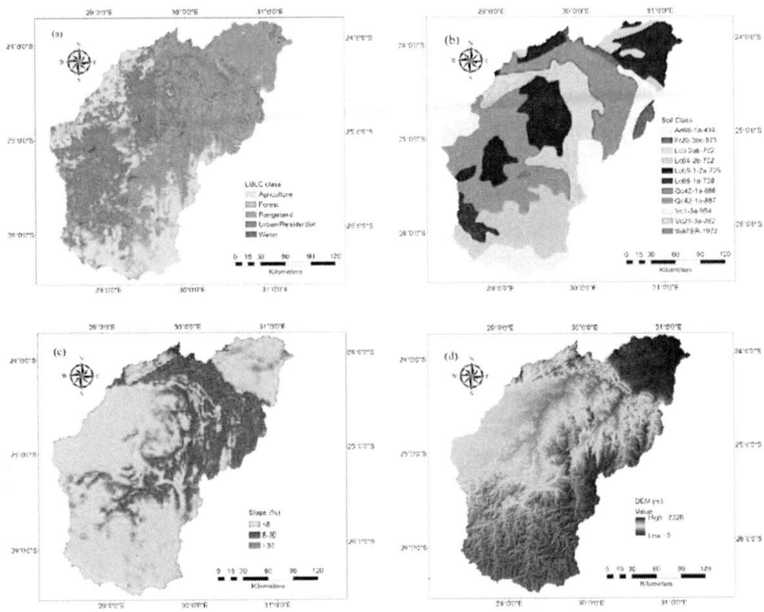

Figure 4. Watershed discretized parameters. (a) Land use, (b) Soils, (c) Slopes and (d) Digital elevation model.

Table 3. Characteristics of the modeled Olifants Basin.

Parameter	Type/class	Area (km²)*	Basin area (%)*
Land use	Agricultural	18446.31	36.88
	Range-Brush	20035.27	40.05
	Residential-High	11313.09	22.62
	Forest-Mixed	228.88	0.46
Soils	Chromic luvisols	19416.50	38.81
	Cambic arenosols	16518.62	33.03
	Chromic vertisols	10611.95	21.21
	Orthic acrisols	2884.26	5.77
	Rhodic ferralsols	592.22	1.18
Slope (%)	<8	38303.66	76.57
	8 - 30	9134.71	18.26
	>30	2585.18	5.17

*Statistics based on threshold values of 10%, 10% and 10% for land use, soils and slope respectively.

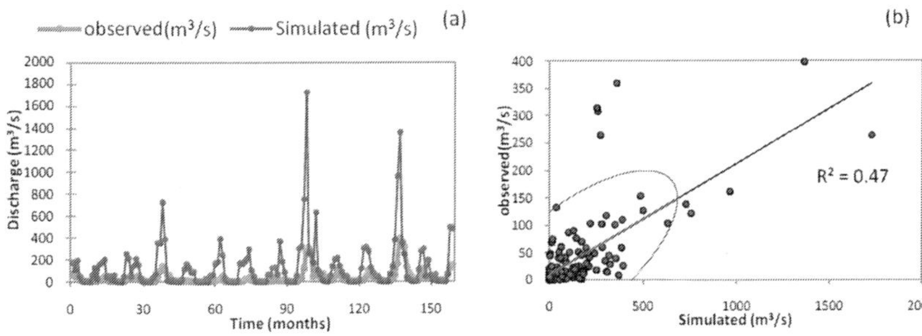

Figure 5. Comparison of simulated to observed monthly discharge (a) discharge hydrograph for 1988-2001 and (b) scatter plots for 1988-2001 at the 95% confidence level.

Although generally, the observed and simulated flow peaked in January, some high peaked predictions were observed in the months of February 1991, 1996 and 2000. An analysis into the correlation between the simulated and observed discharge revealed a coefficient of determination (R^2) of 0.47 (Figure 5(b)) and an NS value of −8.97. Judging the initial model run by the efficiency measures of R^2 and NS it was concluded that the model needed to be calibrated to achieve better parameter estimation required for a better model simulation. Prior to model calibration, relative parameter sensitivity analysis using SUFI-2 algorithm in identifying the most influential parameters on stream flow was carried out.

3.3. Parameter Sensitivity Analysis Using SUFI-2 Algorithm

The results of the relative sensitivity analysis of parameters using SUFI-2 is as shown in Table 4. From the initial fourteen parameters used in the sensitivity analysis, six were found to be sensitive to stream flow based on the t-stat and p-value ($p < 0.05$). Sensitivity of a parameter is determined by the t-stat and how significant that parameter is to the sensitivity analysis is measured by the p-value. Therefore a parameter with larger t-stat and smaller p-value exhibited greater sensitivity on the variable (stream flow) under consideration. Based on this, SCS runoff curve number (CN2) was ranked as the most sensitive parameter on stream flow. Since this parameter is a function of land use, soil permeability and antecedent soil conditions, a change in land use will commensurate into a relative change in stream flow. Other sensitive parameters with relative sensitivity value ranges of 0.00 to 0.02 were the base flow alpha factor for bank storage (ALPHA_BNK), soil evaporation compensation factor (ESCO), soil available water capacity (SOIL_AWC, mm H_2O/mm soil), groundwater delay (GW_DELAY, days) and groundwater "revap" coefficient (GW_REVAP).

Although the following parameters; threshold depth of water in the shallow aquifer required for return flow to occur (GWQMN, mm H_2O), deep aquifer percolation fraction (RCHRG_DP), soil depth (SOL_Z, mm), soil conductivity (SOL_K, mm/h), effective hydraulic conductivity in the main channel (CH_K2, mm/h), base flow alpha factor (ALPHA_BF, days) were included in the sensitivity analysis, it was detected that they were statistically insignificant ($p > 0.05$) and hence a change in their input resulted in no significant effect on stream flow. Findings on parameter sensitivity to stream flow in this study is consistent with other studies around the globe [8] [40] [41] . However, in other jurisdictions parameters such as threshold depth of water in the shallow aquifer required for return flow to occur (GWQMN, mm H_2O), deep aquifer percolation fraction (RCHRG_DP), soil depth (SOL_Z, mm), soil conductivity (SOL_K, mm/h), effective hydraulic conductivity in the main channel (CH_K2, mm/h), base flow alpha factor (ALPHA_BF, days) were detected to be sensitive to stream flow. The variations observed in parameter sensitivity to stream flow can best be explained by the differences inherent in the geomorphological characteristics of the area under investigation.

3.4. Base Flow Separation Prior to Calibration

In order to assess the accuracy of model simulations with regards to base flow index (BFI) estimation, an approach of base flow separation was carried out using observed stream flow data. This provided the basis for acceptance of the simulated base flow index (BFI). The separation of stream flow into base flow and direct runoff components (Figure 6) using the Web-based Hydrograph Analysis Tool (WHAT) indicated substantial amount (67%) of the stream flow to be contributed by direct runoff. BFI from observed stream flow was estimated to be 0.33 which implies a proportion of 33% of stream flow is contributed by base flow. A comparison of the simulated BFI and that obtained using base flow separation from observed data revealed no much significant difference with simulated BFI of 0.24. A breakdown of base flow on yearly basis revealed 66.9% of the total base flow for the calibration period (1988-2001) to have occurred within the period 1997-2001. The earlier years of the calibration period (1988-1996) recorded less base flow of 33.1%.

Table 4. Sensitivity ranking of parameters for modeling stream flow in the Olifants Basin.

Parameter	*t*-stat (absolute values)	*p*-value	Sensitivity ranking
CN2	37.72	0.00	1
ALPHA_BNK	6.97	0.00	2
ESCO	5.57	0.00	3
SOIL_AWC	4.13	0.00	4
GW_DELAY	3.02	0.00	5
GW_REVAP	2.34	0.02	6

1—most sensitive; 6—less sensitive.

3.5. Calibration and Validation of Model

The initial ranges and final calibrated sensitive parameters to stream flow are presented in Table 5. A plot of monthly simulated to monthly observed stream flow (Figure 7 and Figure 8) indicated a good agreement between the datasets with an NSE of 0.88 and R^2 of 0.89 for the calibration period (1988-2001) and NSE of 0.67 and R^2 of 0.78 for the validation period (2002-2013). The model captured well the monthly time series of stream flow as well as the trend during both calibration and validation periods.

Although, observed and simulated discharge matched well for calibration and validation periods there was an average magnitude overestimation of the observed discharge with PBIAS of −11.49% for the calibration period and −20.69% for the validation period. The observed and simulated mean annual flow (MAF) during the calibration period (1988-2001) was 1175 Mm^3/a (37.27 m^3/s) and 1310 Mm^3/a (41.55 m^3/s) respectively which represent an overestimation of the observed stream flow by 11.48%. For the validation period (2002-2013) the observed and simulated MAF were 1193 Mm^3/a (37.82 m^3/s) and 1440 Mm^3/a (45.65 m^3/s) respectively also indicating an overestimation of the observed by 20.69%. The cumulative simulated MAF of 2750 Mm^3/a achieved in this study compares differently from the 2040 Mm^3/a and 1992 Mm^3/a recorded by the WSAM [24] and WR90 [42] studies respectively for the Basin. The difference may be attributed to the spatio-temporal change in rainfall occurring throughout the years. Considering the model performance statistics (Table 6) for both calibration and validation periods and on the criteria set out by Moriasi et al. [39] for assessing model performance, the set up model was rated "good" for the Olifants Basin. These measures of model performance indicate that the model captures well the observed stream flow.

However, a comparison of the measures of statistics for the calibration and validation period (Table 6) reveals a better performance of the model during the calibration period as compared to the validation period. This observation agrees with the studies of Moriasi et al. [39] in which model performance during the calibration period did well compared to the validation period.

4. CONCLUSIONS AND RECOMMENDATION

Demonstrated in this study is the calibration, validation and uncertainty analysis for the Olifants Basin. The calibrated SWAT model indicated statistical reliability in simulating the stream flow of the Olifants Basin. However a comparison of the model during calibration and validation revealed a better simulation at the calibration stage than at the validation stage. This is to suggest that if data are available for other variables (i.e. sediment, water quality, etc.) the model can equally be calibrated to simulate those variables with good measures of statistics. With available and reliable data, the SWAT tool comes in handy as a decision support tool that can influence policy directions on watershed management processes and more especially on water resources.

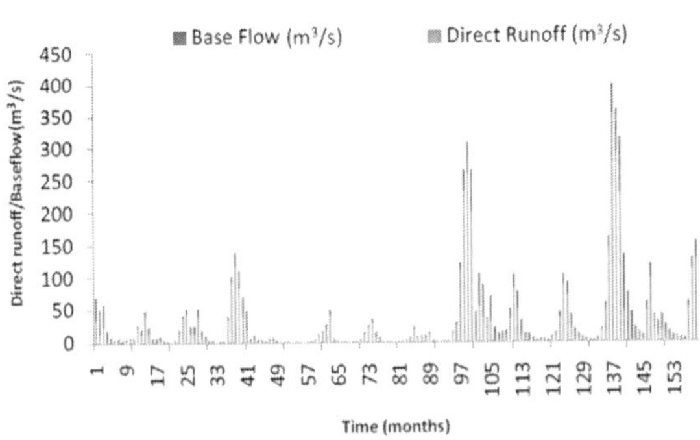

Figure 6. Monthly base flow separation for the calibration period 1988-2001.

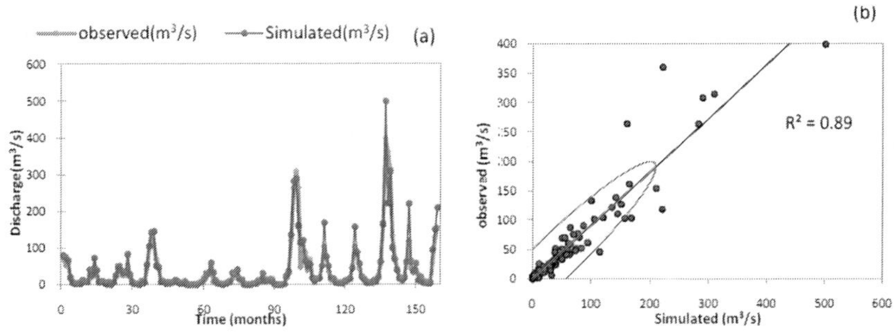

Figure 7. Monthly discharge calibration for the period 1988-2001, (a) hydrograph analysis and (b) scatter plot at the 95% confidence level.

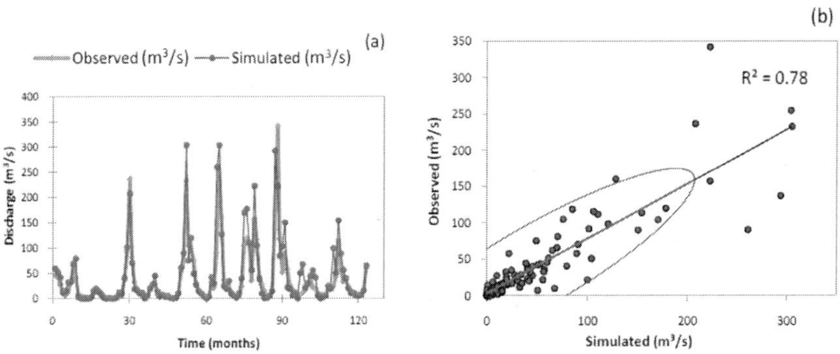

Figure 8. Monthly discharge validation for the period 2002-2013, (a) hydrograph analysis and (b) scatter plot at the 95% confidence level.

Table 5. Model calibrated parameter values.

Parameter	Description	Range	Fitted value
CN2	SCS runoff curve number	35 - 98	65*
ALPHA_BNK	Base flow alpha factor for bank storage	0 - 1	0.39
ESCO	Soil evaporation compensation factor	0 - 1	0.67
SOIL_AWC	Soil available water capacity	0 - 1	0.2
GW_DELAY	Groundwater delay (days)	0 - 500	345
GW_REVAP	Groundwater "revap" coefficient	0.02 - 0.2	0.15

*Average estimated for the basin.

Table 6. Monthly time step calibration and validation performance statistics.

Model stage	Evaluation statistics			
	NSE	R^2	PBIAS (%)	RSR
Calibration (1988-2001)	0.88	0.89	−11.49	0.34
Validation (2002-2013)	0.67	0.78	−20.69	0.57

ACKNOWLEDGEMENTS

Tshwane University of Technology is acknowledged for the financial support for the study. The authors express their gratitude to the Department of Water Affairs and the South African Weather Service for freely providing the data used for the study.

REFERENCES

1. Betrie, G.D., Mohamed, Y.A., van Griensven, A. and Srinivasan, R. (2011) Sediment Management Modeling in the Blue Nile Basin Using SWAT Model. Hydrology and Earth System Sciences, 15, 807-818.

2. Panhalkar, S.S. (2014) Hydrological Modeling Using SWAT Model and Geoinformatic Techniques. The Egyptian Journal of Remote Sensing and Space Sciences.

3. Singh, V., Bankar, N., Salunkhe, S.S., Bera, A.K. and Sharma, J.R. (2013) Hydrological Stream Flow Modeling on Tungabhadra Catchment: Parameterization and Uncertainty Analysis Using SWAT CUP. Current Science, 104, 1187-1199.

4. Adeogun, A.G., Sule, B.F. and Salami, A.W. (2015) Simulation of Sediment Yield at the Upstream Watershed of Jebba Lake in Nigeria Using SWAT Model. Malaysian Journal of Civil Engineering, 27, 25-40.

5. DWAF (2005) Olifant River Water Resources Development (ORWRP) Environmental Impact Assessment (12/12/20/ 553). Infrastucture Development.

6. Obuobie, E. and Bernd, D. (2008) Using SWAT to Evaluate Climate Change Impact on Water Resources in the White Volta River Basin. West Africa Conference on International Research on Food Security, Natural Resource Management and Rural Development, University of Hohenheim, Tropentag, 7-9 October 2008.

7. Merrey, D.J., Drechsel, P., Penning de Vries, F.W.T. and Sally, H. (2005) Integrating "Livelihoods" into Integrated Water Resources Management: Taking the Integration Paradigm to Its Logical Next Step for Developing Countries. Regional Environmental Change, 5, 197-204.

8. Setegn, S.G., Srinivasan, R. and Dargahi, B. (2008) drological Modeling in the Lake Tana Basin, Ethiopia Using SWAT Model. The Open Hydrology Journal, 2, 49-62.

9. Chou, S., Russ, B. and Anne, W. (1997) Water Scarcity in River Basins as a Security Problem. Woodrow Wilson International Centre for Scholars, Environmental change and Security Project, Washington DC, 96-105.

10. Xie, M. (2006) Integrated Water Resources Management (IWRM)—Introduction to Principles and Practices. World Bank Institute.

11. Carius, A., Dabelko, D.G. and Wolf, A.T. (2004) Water, Conflict and Cooperation, Policy Briefing Paper for United Nations and Global Security Initiative.

12. UNDP (2006) Summary Human Development Report. Beyond Scarcity: Power, Poverty and the Global Water Crisis.

13. Biswas, A.K. (2001) Management of International Waters: Opportunities and Constraints, in Sustainable Development of the Ganges-Brahmaputra-Meghna Basins. In: Biswas, A.K. and Uitto, J.I., Eds., United Nations University Press, Tokyo, 1-16.

14. Gleick, P.H. (2000) The Changing Water Paradigm: A Look at Twenty-First Century Water Resources Development. Water International, 25, 127-138.

15. Ullrich, A. and Volk, M. (2009) Application of the Soil and Water Assessment Tool (SWAT) to Predict the Impact of Alternative Management Practices on Water Quality and Quantity. Agricultural Water Management, 96, 1207-1217.

16. Schuol, J., Abbaspour, K.C., Srinivasan, R. and Yang, H. (2008) Estimation of Freshwater Availability in the West African Sub-Continent Using the SWAT Hydrologic Model. Journal of Hydrology, 352, 30-49.

17. Yang, J., Reichert, P. and Abbaspour, K.C. (2007) Bayesian Uncertainty Analysis in the Distributed Hydrologic Modeling: A Case Study in the Thur River Basin (Switzerland). Water Resources Research, 43, W10401.

18. Simic, Z., Milivojevic, N., Prodanovic, D., Milivojevic, V. and Perovic, N. (2009) SWAT-Based Runoff Modeling in Complex Catchment Areas—Theoretical Background and Numerical Procedures. Journal of the Serbian Society for Computational Mechanics, 3, 38-63.

19. Neitsch, S.L., Arnold, J.G., Kiniry, J.R. and Williams. J.R (2005) Soil and Water Assessment Tool Theoretical Documentation. Version 2005.

20. Arnold, J.G. and Fohrer, N. (2005) SWAT2000: Current Capabilities and Research Opportunities in Applied Watershed Modeling. Hydrological Processes, 19, 563-572.

21. Obuobie, E. and Diekkrüger, B. (2008) Using SWAT to Evaluate Climate Change Impact on Water Resources in the Olifant River Basin, West Africa. Conference on International Research on Food Security, Natural Resource Management and Rural Development, University of Hohenheim, Stuttgart, 7-9 October 2008.

22. Li, Z., Liu, W., Zhang, X. and Zheng, F. (2009) Impacts of Land Use Change and Climate Variability on Hydrology in an Agricultural Catchment on the Loess Plateau of China. Journal of Hydrology, 377, 35-42.

23. Neitsch, S.L., Arnold, J.G., Kiniry, J.R., Williams, J.R. and King, K.W. (2011) Soil and Water Assessment Tool Theoretical Documentation Version 2009. Grassland, Soil and Research Service, Temple.

24. Schultz, C.B. and Watson, M.D. (2002) WSAM: Water Situation Assessment Model Version 3: A Decision Support System for Reconnaissance Level Planning. Volume 1: Theoretical Guide. ARCUS-GIBB and DWAF, Republic of South Africa.

25. IWMI (2008) Baseline Report Olifants River Basin in South Africa: A Contribution to the Challenge Program Project 17 "Integrated Water Resource Management for Improved Rural Livelihoods: Managing Risk,

Mitigating Drought and Improving Water Productivity in the Water Scarce Limpopo Basin".

26. Schulze, R.E., Maharaj, M., Lynch, S.D., Howe, B.J. and Melvil-thomson, B. (1997) South African Atlas for Agrohydrology and Climatology. University of Natal, Pietermaritzburg.

27. FAO (2002) Major Soils of the World. Land and Water Digital Media Series. Food and Agricultural Organization of the United Nations. FAO, Rome.

28. FAO (2005) Digital Soil Map of the World and Derived Soil Properties of the World. Food and Agricultural Organization of the United Nations. FAO, Rome.

29. Batjes, N.H. (2002) ISRIC-WISE Global Data Set of Derived Soil Properties on a 0.5 by 0.5 Degree Grid (Version 2.0). Report 2002/03. International Soil Reference and Information Centre (ISRIC), Wageningen.

30. Anderson, J.R., Hardy, E.E., Roach, J.T. and Witmer, W.E. (1976) A Land Use and Land Cover Classification System for Use with Remote Sensing Data. USGS Professional Paper 964, 138-145.

31. Neitsch, S.L., Arnold, J.G., Kiniry, J.R., Srinivasan, R. and William, J.R. (2002) Soil and Water Assessment Tool User's Manual. Version 2000.

32. FAO (2003) The Digital Soil Map of the World. Version 3.6.

33. Abbaspour, K.C. (2012) SWAT-CUP-2012.SWAT Calibration and Uncertainty program—A User Manual. Swiss Federal Institute of Aquatic Science and Technology, Dübendorf.

34. Faramarzi, M., Abbaspour, K.C., Schulin, R. and Yang, H. (2009) Modelling Blue and Green Water Resources Availability in Iran. Hydrological Processes, 23, 486-501.

35. Schuol, J., Abbaspour, K.C., Yang, H., Srinivasan, R. and Zehnder, A.J.B. (2008) Modeling Blue and Green Water Availability in Africa. Water Resources Research, 44, W07406.

36. Abbaspour, K.C., Yang, J., Maximov, I., Siber, R., Bogner, K., Mieleitner, J., Zobrist, J. and Srinivasan, R. (2007) Modelling Hydrology and Water

Quality in the Pre-Alpine/Alpine Thur Watershed Using SWAT. Journal of Hydrology, 333, 413-430.

37. Abbaspour, K.C., Johnson, C.A. and van Genuchten, M.T. (2004) Estimating Uncertain Flow and Transport Parameters Using a Sequential Uncertainty Fitting Procedure. Vadose Zone Journal, 3, 1340-1352.

38. Coffey, A.E., Workman, S.R., Taraba, J.L. and Fogle, A.W. (2004) Statistical Procedures for Evaluating Daily and Monthly Hydrologic Model Predictions. Transaction of the American Society of Agriculture Engineers, 47, 59-68.

39. Moriasi, D.N., Arnold, J.G., van Liew, M.W., Bingner, R.L., Harmel, R.D. and Veith, T.L. (2007) Model Evaluation Guidelines for Systematic Quantification of Accuracy in Watershed Simulations. Transactions of the ASABE, 50, 885-900.

40. Vaghefi, S.A., Mousavi, S.J., Abbaspour, K.C., Srinivasan, R. and Yang, H. (2013) Analyses of the Impact of Climate Change on Water Resources Components, Drought and Wheat Yield in a Semi-Arid Regions: Karkheh River Basin in Iran. Hydrological Processes, 28, 2018-2032.

41. Yang, H., Reichert, P., Abbaspour, K.C., Xia, J. and Yang, H. (2008) Comparing Uncertainty Analysis Techniques for SWAT Application to the Chaohe Basin in China. Journal of Hydrology, 358, 1-23.

42. Midgley, D.C., Pitman, W.V. and Middleton, B.J. (1994) Surface Water Resources of South Africa, Volumes I, II, III, IV, V and VI. Reports No's. 298/1.1/94, 298/2.1/94, 298/3.1/94, 298/4.1/94, 298/5.1/94 and 298/ 6.1/94, Water Research Commission, Pretoria.

Chapter 10

Traditional Lowlands Water Management in Dano, South-Western Burkina Faso

Sié Pale[1*], Dapola Evariste Constant DA[2]

[1]Polytechnic University Center of Ouahigouya/Department of Atmosphere Sciences

[2]University Ouaga I, Pr Joseph KI-ZERBO/Department of geography

ABSTRACT

Lowland water resources management represents a challenge of the future that calls any community. Irrigated crops are grown in some areas of Burkina Faso, others are limited by a lack of irrigation infrastructure. Due to limited crop irrigation, crops and the associated populations dependent on them, depend on rain and on climatic factors. Thus, there is a need to understand and implement traditional mechanisms for managing lowland water in Dano, where climatic and geological conditions provide a sustained source of water. Here, I use a literature review combined with field work and interviews/questionnaires to estimate the potential exploitable plains to 16,056 ha or 24% of the communal area. Management mechanisms and traditional operating systems of lowland waters were clear, which helped to set the technological level of farmers, in partial control of water management.

KEYWORDS

Burkina Faso, Dano, South Sudanese Climate, Lowlands, Traditional Management

1. INTRODUCTION

Traditional agriculture has long fed the rural population, however, erratic rainfall since the 1970s, degradation of natural resources (soil, water, vegetation) due to population pressure have contributed to decreased agricultural yields [1] [2] .

Water deficiency due to climatic and anthropogenic factors and rapid population growth has reduced the availability of water and land per capita; it is one of the main factors responsible for this evolution which is explained by the decrease of agricultural production and forest resources [3] . Meanwhile, municipal needs facing limited water resources have been increasing. This shows how it is essential to maintain the rural development programs, geared towards agriculture, particularly in sub-Saharan Africa. These programs have an obligation to simultaneously increase productivity, reduce poverty and improve gender equity; these are three factors that seem decisive for the improvement of food security [2] - [4] .

The water supply off season for small productive activities such as gardening, fruit growing and nursery of exotic plants, helps to diversify agricultural activities and to cope with shortages of arable land [5] - [7] . Irrigation of small agricultural areas uses a variety of techniques. These have become more efficient in recent decades. This irrigation enables farmers to take advantage of the water available in rivers or aquifers to extend the production time, adding to traditional rainfall, those of dry season crops (contresaison) [8] - [11] .

In Burkina Faso, the droughts of the 1970s and 1980s still mark deeply the lives of rural populations; they have led to environmental degradation, population movements towards the wetter areas and large cities, as well as a situation of continuing food insecurity [12] . These droughts are partly responsible for the impoverishment of populations, victims of a serious agro-ecological crisis and generate the progressive disintegration of traditional societies. The relationships between populations in dry regions and those irrigated areas are inevitably changed, evolving from a certain complementarity to forms of dependence [2] [13] [14] .

With more than 2000 dams including 300 permanent and 8 large dams [15] [16] completed, often in earth and gabion baskets, and equipped with small irrigated areas downstream, Burkina Faso is a model for other Sahel-Sudanese countries. These dams are the primary means of mobilizing surface water not only for agriculture but also for drinking water needs, animal husbandry and industry. Lowlands have been the target of many small improvements (filtering dikes, rice sills spilling). Access to these improved areas is difficult and often, the cause of exclusion of dominated social groups without land guarantees; these are notably women, newcomers, pastoralists and cadets [17] - [20] .

The Water Resources development program in the South-west (Ressources en Eau dans le Sud-ouest: RESO) considered sustainable management with a blueprint and concerted planning, forming the framework of development stakeholders; it supports local initiatives through sub-programs focusing on assessment and recovery of water resources. Careful monitoring of innovations is made to conduct a consultation among stakeholders and empower them, by setting up agencies, basin Committees and Local Committees for Water (Comité Local de l'Eau: CLE). However, a general regulation blur is a handicap, combined with a lack of references and weak capacity of farmers. The integrated management approach does not yet seem quite understood by the operators [18] [21] [22] .

If irrigated crops are possible in some areas of Burkina Faso, others suffer more from lack of irrigation infrastructure capable to regulate the problem of rural employment. In the case of Kou valley example, farming systems of the populations are mostly organized in strict rainfall crops, in lowland and small irrigation cultures practiced on developed areas and other so-called peripheral informal perimeters. These systems are weakly intensified, undiversified and offer a low productivity [23] . Now pushed toward the lowlands, previously unexploited areas [24] , populations are still mainly engaged in traditional crops depending on climatic factors, particularly rainfall [25] . It is in this context that we have to know, how are the lowlands traditionally managed in Dano?

The general objective of this study is to explain the traditional lowland management mechanisms in Dano. What is the importance of agricultural lowlands of Dano Commune which occupies a favorable position in the climate area in terms of rainfall amounts? How is it managed in the current situation of natural resource degradation, coupled with the decline in rainfall and population pressure? At what level of water control the lowlands traditional management can be located?

2. METHODOLOGY

The approach uses the traditional data collection techniques, taking into account mainly the literature review and primary data collection. This approach is described on the following points:

2.1. The Study Site

Dano commune is located between 10°58'01" and 11°19'28" N and, 2°49'48" and 3°11'20" W. Capital of the Province of Ioba, Dano covers an area of 669 km^2 and has 22 administrative villages. It is located at a distance of 280 km from Ouagadougou [26] [27] (see Figure 1).

Climate field of Dano is South Sudan type characterized by the alternation of two distinct seasons (dry season that lasts six to seven months). The average annual rainfall exceeds 900 mm.

The town of Dano is home to a variety of ethnic groups. The majority Dagara, coexist with Pougouli, Bwaba, Mossi and Fulani. Each of these ethnic groups, while following the dictates of his social organization, respects the customs and traditions of the other [26] .

Agriculture is the main economic activity. It occupies more than 90% of the population. Agricultural production is mainly food-type. Indeed, during the 2003-2013 period, food crops (sorghum, millet, maize and rice) have held an average of 65% of planted areas against 34% for cash crops. Apart from rainfall, is practiced in the dry season, market gardening and growing small irrigation [28] [29] . Vegetable crops are grown in the lowland Lofing, around the Moutori dam and Pontièba water storage [26] .

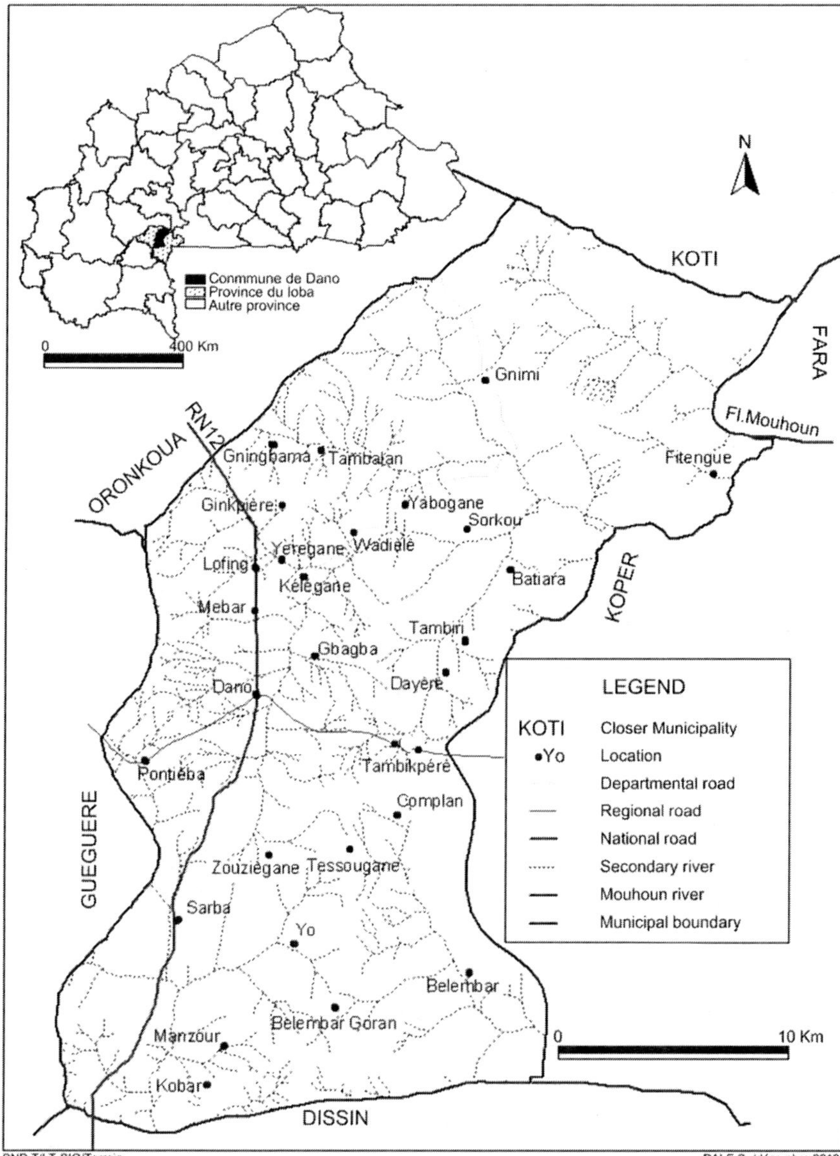

Figure 1. The town and surrounding Dano.

2.2. The Literature Review

In the literature, the channels used are the websites and documentation centers. Quantitative and qualitative data about study were collected through technical reports of the Ministry of Agriculture and non-governmental organizations, student dissertations and scientific articles. The relevance of the topics covered by the different authors and their areas of investigation were taken into account for the synthesis documents.

2.3. The Primary Data Collection

A survey was conducted among 96 lowland farmers in Dano Commune, particularly in sectors 3, 6 and 7 (Pontièba). The technique used is simple random sampling, without any statistical law. Interviews were also conducted with resource persons of services devolved to Dano such as the Provincial Director of Agriculture, Hydraulics and Fisheries, the Provincial Director of Environment and Sustainable Development, the municipality of Dano and the Dreyer Foundation. The collection tools used are designed to provide quantitative and qualitative data.

2.4. Processing, Data Analysis and Presentation of Results

Data from surveys and interviews are grouped according to our secondary research questions and the presentation of results in three parts. The first analysis allowed us to bring out an inventory of hydro-agricultural infrastructure in the form of a map using the Geographical Information Systems (GIS). Then, mechanisms and traditional management systems of the lowlands are mentioned and finally, an analysis is made followed by discussions to locate the degree of technological implication of this rural population particularly in lowlands water resources management.

3. RESULTS AND DISCUSSION

The results obtained follow the implementation of a methodological approach described below:

3.1. Areas of Lowland

The integration of rice as a cereal camp enabled local people to briefly occupy the floodplains. The BUNA- SOLS's work in 2000 [30] have established agronomic abilities of soils for the main cereals produced in Dano, including units capable of providing double cropping, rainfall and dry crops, as shown in Table 1.

By observing this table, agronomic units capable of supporting rice cultivation, which most often occupy flood areas or slums, are the units 1, 2 and 3 having an average ability (S_{2n}). It's all ferruginous leached hydromorphic soils, leached ferruginous soils on indurated inclusion, waterlogged soil surface pseudogley and slightly evolved hydromorphic alluvial soils intake. The area they occupy equivalent to 24% of the total area of the municipality [27] (see Figure 2). These soil types also support the culture of other cereals mentioned in the above table. The other soil types, unit 3 consists of hydromorphic soils low in humus surface pseudogley (S_{3W}); unit 4 of leached tropical ferruginous soils moderately deep indurated and indurated deep leached tropical ferruginous soils (S_{3r}); unit 5, unfit for any cereal production (N_2), includes lithosols on rock, lithosols on ironstone, leached superficial indurated ferruginous tropical soils and indurated ferruginous leached deep soils [27] [31] [32] .

Compared to the density of the river system, the proportion of lowland soils appears low because of the terrain. As a reminder, hills, mounds and glazes dominate the landscape Dano. However, note that 24% of potential land for rice cultivation and/or contresaison (dry crops) is not negligible. Expressed in hectares, this percentage amounts to 16,056 ha, or about 7% of the national potential [33] .

So far, four sites are involved in vegetable production in the dry season; the lowland Lofing, withholding water of Ligmarè and Pontièba and the Moutori landscaped perimeter. It is the latter that is most important with 23 ha of exploiting area [31] . Except the area of Ligmarè which is purely operated traditionally, others experienced an arrangement with irrigation channels in Pontièba and Moutori, and culvert wells at Lofing.

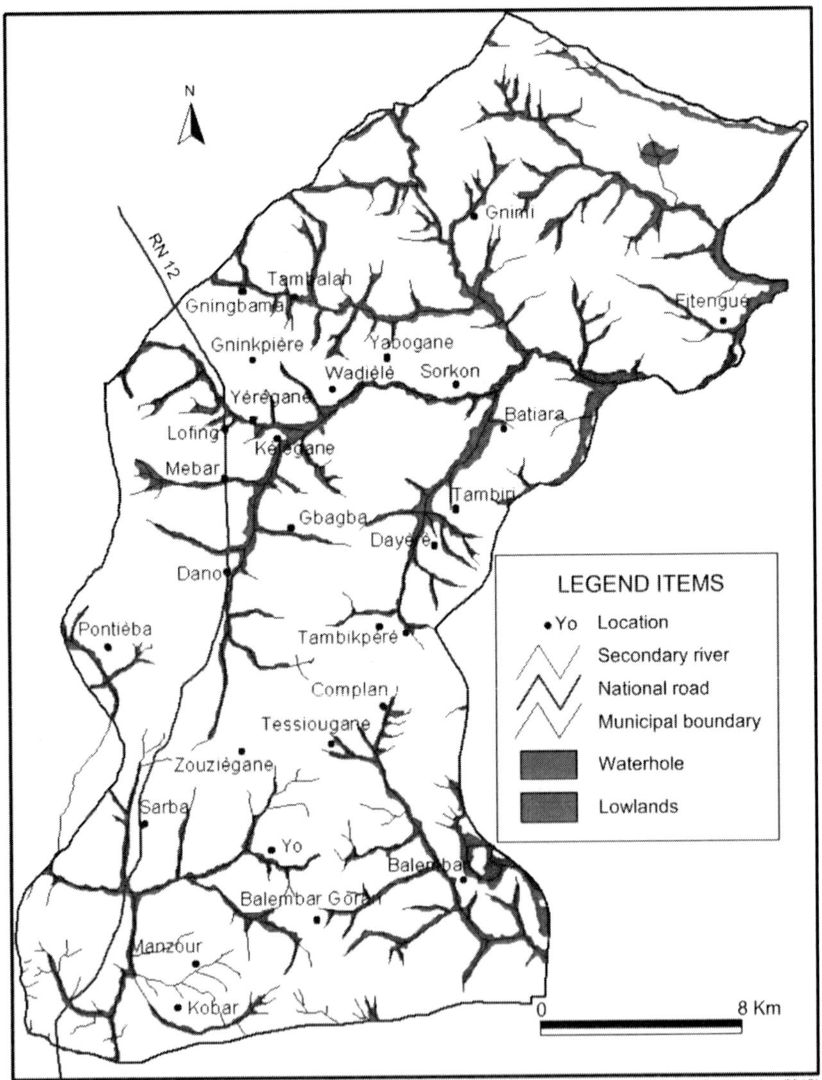

Figure 2. The extent of lowlands in Dano.

Table 1. Ability soil agronomic units for grain Dano.

Agronomic Units	Proportion (%)	Maize	Millet	Sorghum	Rice
1	10	S_{2n}^{*}	S_{2n}	S_{2n}	S_{2n}
2	1	S_{2n}	S_{2n}	S_{2n}	S_{2n}
3	13	S_{3w}^{**}	S_{3w}	S_{3w}	S_{2n}
4	35	S_{3r}	S_{3r}^{***}	S_{3r}	N_2
5	41	N_2^{****}	N_2	N_2	N_2

Source: Based AGRITECH FASO (2010).
Please note: $^{*}S_{2n}$ = average ability; $^{**}S_{3W}$/$^{***}S_{3r}$ = marginal ability; $^{****}N_2$ = permanently incapacitated.

3.2. Traditional Management Mechanisms of Water Resources

The right to water is closely linked to that of the land which is acquired by the principle of the precedence or seniority. And control of traditional water management (TWM) systems (Gestion Traditionnelle de l'Eau: GTE) is provided by traditional leaders who are indigenous. These TWM or GTE (in French) are governed by customary law with a village ownership [34] . Only holders' customary rights generally inherited through the oral tradition. Recently, SOME J-M [35] mentions that on the earth, the heirs are fighting to avoid spoliation operated by the state or foreign powers. Their struggle, which is not just only a defense of any property, takes the form of a crusade to ensure that land managers who received this "gift of the earth" know their rights, while remaining open to trade necessitated by (and for) the development [36] - [38] .

The land crops in the lowlands are obtained by inheritance father to son, nephews and cousins. In the community, no woman can be hold land. According to the survey results, the powers of such land plots, that is to say to foreign applicants to the family, are only temporary and the owners have the right to demand it at any time [36] .

The exploitation of surface waters in the lowlands is allowed for all inhabitants of the land. In the dry season, the water in the troughs is in the form of beads; in places, wells are dug to overcome the problem of watering animals and gardening. Furthermore, the animals must be kept away from existing fields, to avoid internal disputes within the village. To this end, corridors of access to water points and grazing areas are drawn.

3.3. The Traditional Lowlands Operating Systems

In the lowlands, farmers traditionally sown early in the rainy season after summary farming methods performed with the daba. The first floods come as the seedlings are large enough to support them [39] . Compared to the farmers of developed areas, those who practice purely rain-based watering, recognize that they are taking longer to produce or grow than previous. The plots are prepared before or during the first rains; it is a simple plowing by the hoe or large mounds of at least 2 square meters of 30 to 50 cm in height and spaced from each other by at least 1 m. These mounds can be made in rows or staggered; farmers prefer the latter provision to slow water currents capable of carrying some mounds.

Seedlings also follow a certain logic (as a reminder, all farmers practice mixed farming in the operating system); on mounds, corn is planted, often accompanied okra around them. Some mounds may also harbor Cucurbitaceae (gourd family), especially the squash to be harvested before the floods. The spaces between the mounds are exclusively rice (long cycle, traditional, 4 - 5 months). To slow runoff and erosion, seedlings are usually arranged in staggered rows (see Figure 3(a) and Figure 3(b)).

In the case of gardening, which lasts 3 - 4 months from November to February, the boards are inter-spaced not more than 50 cm. Common vegetables such as tomatoes, eggplants, okra, peppers, cabbage and increasingly onion are produced. Traditional operators recognize they have no control in the production of onion. Figure 4 presents a technique traditional operating lowlands to Ligmarè.

3.4. The Degree of Technological Involvement of Farmers in Traditional Management of Water Resources

Water and land are seen as divine gifts, often representing a relationship between God, genies, and human beings; they are sacred elements related to the ordering in different societies. Water resources control forms are always determined by this design [40] . The need to fill a food deficit facing erratic rainfall and a population explosion that compresses the cultural space, led farmers to focus more on floodplains [41] . Lowlands in Dano, formerly used for growing tobacco, pepper or eggplant are nowadays exploited for cultivation of cereals, including maize and rice. This enthusiasm comes from the strong climate variability that becomes elusive for farmers. The 2012 IUCN report on wetlands, with reference to flood areas or lowlands, particularly in France, confirms this argument; for her, wetlands have also long been seen as detrimental ecological environments, which would explain the many improvements made over time. Nowadays, these wetlands regress due mainly drainage for agricultural purposes [24] . Thus, rice production is growing for a decade and forcing rural people to a new civilization with water [42] - [44] . The three types of production such as irrigated rice, rice lowland and upland strict that are practiced throughout the town. Nationally, the Ioba Province is one of provinces occupying the 3rd rank rice production [45] .

Rice production in Dano has become a booming business and the degree of involvement of farmers in water management is increasingly important. The peasant effort is laudable even if the majority does not have total control of agricultural water. The gradual intensification of agriculture requires more efficient water management. And a key to the problem of water for the rural world is the creation and multiplication in recent years, developed areas [46] . The results show the existence of three types of lowland occupied by farmers. First, traditionally exploited lowlands are most prevalent; then the simple managed lowlands and finally the improved one. It should be noted here that these three groups of lowland could not be quantified. So, the degree of involvement in the peasant agricultural water management is at three levels; respectively between those who have had no training, those who have benefited from training or study tours and those who are permanently assisted by agricultural technicians [47] .

Source: PALE S. (June 2012).

These two pictures show two ways to carry out the mounds in the lowlands and arrangement of crops; we clearly observe the alternation of sowing on the microtopography created by farmers (corn on small hills and rice between them).

(a) (b)

Figure 3.Arrangement of mounds and types of crops in traditional farming to lowlands in Dano.

The main coaching topics for those receiving assistance include the application of technical ideas, composting, respect the cropping calendar and interviews plots. The purpose of the application of this technology package is to enable producers to improve yields in their plot.

Unfortunately, they often face due to insufficient number of breaks support technicians. This inadequacy of the staff hampers the normal monitoring of producers and promotes non-compliance with rules set in different arrangements [39] .

Source: PALE S. (June 2012).

The illustration shows the boards with regular shapes separated from each other by an alley of about 50 cm. They wear tomato plants in the background and those of eggplant forward; it is the garden below the dam of Ligmarè, 15 km from Dano, further south.

Figure 4.Arrangement of boards and cultures within a garden in Ligmarè lowland (South Dano).

At this stage, a tiny fraction of Dano rural population receives training or partial control study trips water. The majority of farmers expressed their enthusiasm to exploit in the irrigated perimeter of Moutori or, at best, to usurp technology. Unfortunately, this technology package seems to monopolize their complex because of the lack of equipment and infrastructure, income poverty and to appropriate it as illiteracy or low educational attainment. This is a commitment, determination, and even an awareness of rural population to move to other technologies to improve crop yields; this deserves special attention of policy makers and NGOs for them.

4. CONCLUSIONS

Geological formations and exogenous agents of erosion have promoted a relief accident in the Dano Commune. It forms a dendritic network of streams rather important in more or less cramped plains. These floodplains reported in the area of the town, occupy 24% or 16,045 ha. These lowlands are traditionally run by local people. And management mechanisms are based on the prior definition of allocation with the involvement of the beneficiaries, when it comes to amenities or are provided by traditional leaders who are indigenous.

Water control is located at three levels according to the three types of lowland identified (purely traditional lowlands, simple managed lowlands and improved lowlands). Thus, it appears that only a small portion of this population has partial control of water.

Attention of policy makers and humanitarian organizations working in the field of rural development, must be paid to this population that expresses a strong enthusiasm to go to other supposedly improved technologies that can increase agricultural production in the storm system and gardener.

REFERENCES

1. Mahé, G., Diello, P., et al. (2010) Declining Rainfall and Increased Flows in the Sahel: Climate and Human Impact on the Nakambe Flows in Burkina Faso [Baisse des pluies et augmentation des écoulements au Sahel: Impact climatique et anthropique sur les écoulements du Nakambe au Burkina Faso]. Sécheresse, 21, 1-6.

2. PNUE (2002) 3rd Joint Regional Meeting of National Coordinators and the Committee “Project of integrated water resources management and Volta Basin lands”, Accra, 9-10 December 2002. Mission Report. [3ème Réunion Régionale conjointe des coordinateurs nationaux et du comité du “Projet de gestion intégrée des ressources en eau et des

terres du bassin de la Volta", Accra, du 9 au 10 décembre. Rapport de Mission], MAHRH/DGIRH, Burkina Faso, 34 p.

3. Bognini, S. (2011) Impacts of Climate Change on Vegetable Crops in Northern Burkina Faso: Case of Ouahigouya [Impacts des changements climatiques sur les cultures maraîchères au nord du Burkina Faso: Cas de Ouahigouya]. RENAF/SMHI, Ouagadougou, 38 p.

4. FAO (2002) Water and Agriculture, Producing More with Less Water [Eau et agriculture, produire plus avec moins d'eau]. FAO/FIAT/PANIS, Rome, 28 p.

5. Dipama, J.-M. (2004) Peasant Perception of Environmental Degradation on the Perimeter of Kompienga Dam (Burkina Faso). In: Priorities and Practices of Rural Development in Sub-Saharan Africa [Perception paysanne de la dégradation de l'environnement sur le pourtour du lac du barrage de la Kompienga (Burkina Faso). In: Priorités et pratiques du développement rural en Afrique subsaharienne.] Univ. Ouagadougou/ IDR/UPB/DANIDA, Ouagadougou, 149-164.

6. FAO (2008) Water and Rural Poverty: Interventions for Improving Livelihoods in sub-Saharan African Populations [Eau et pauvreté rurale: Interventions pour améliorer les moyens d'existence des populations d'Afrique subsaharienne]. FAO/FIDA, Rome, 126 p.

7. Laouina, A., Aderghal, M., Al Karkouri, J., Chaker, M., Machmachi, I., Machouri, N. and Sfa, M. (2010) Land Use, Runoff and Land Degradation: Case of Sehoul Zone, Atlantic Region (Morocco) [Utilisation des sols, ruissellement et dégradation des terres: Le cas du secteur Sehoul, région atlantique (Maroc)]. Sécheresse, 21, 309-317.

8. Bouaicha, R. and Benabdelfadel, A. (2010) Variability and Surface Water Management in Morocco [Variabilité et gestion des eaux de surface au Maroc]. Sécheresse, 21, 1-5.

9. Cessouma, A. (2008) Irrigated Agriculture and Water Saving Opportunities in Mali [L'agriculture irriguée et les perspectives d'économie d'eau au Mali]. Revue HTE No.141, décembre, Rabat/ Maroc, 4-6.

10. Chahbani, B. (2004) Technical Innovations for Optimizing the Collection, Conservation and Water Use for Sustainable Development of Rainfall Agriculture in Arid Regions [Innovations techniques pour

l'optimisation de la collecte, de la conservation et de l'utilisation de l'eau pour un développement durable de l'agriculture pluviale dans les régions arides]. Institut des Régions Arides, Tunisie, 73-78.

11. El Garouani, A. and Merzouk, A. (2010) Geomatics Approach for Defining Protection Areas for Groundwater Resources (Well-Field) [Approche géomatique pour la délimitation des périmètres de protection des ressources en eau souterraines (champ captant)]. http://www.secheresse.info/spip.php?article1952

12. Touaibia, B. (2010) Problems of Erosion and Sediment Transport in Northern Algeria [Problématique de l'érosion et du transport solide en Algérie septentrionale]. Sécheresse, 21, 1-6.

13. Brondeau, F. (2005) Water Access, Differentiation Factor of Landscapes and Rural Societies: e.g. Irrigated Perimeters of the Office du Niger and Their Dry Borders [L'accès à l'eau, facteur de différenciation des paysages et des sociétés rurales: exemple des périmètres irrigués de l'Office du Niger et de leurs marges sèches]. http://www. secheresse. info/spip.php?article1817

14. MAHRH (2002) Action Plan for Integrated Management of Water Resources (IWRM) (2003-2015). Volume 1: Strategic Plan of IWRM actions, 3rd version [Plan d'Action de Gestion Intégrée des Ressources en Eau (PAGIRE) (2003-2015). Tome 1: Plan d'actions stratégique de la GIRE, 3ème version]. DGIRH/DANIDA, Ouagadougou/Burkina Faso, 52 p.

15. MEE/GIRE (2001) Inventory of Water Resources in Burkina Faso and Their Context Management. Final Version [Etat des lieux des ressources en eau du Burkina Faso et de leur cadre de gestion. Version finale]. MEE-SP, GIRE—May, Ouagadougou, 243 p.

16. REEB (2002) Report on environment in Burkina Faso, Summury [Rapport sur l'état de l'environnement au Burkina Faso, Résumé]. Sp/CONAGESE, 1rst Edition, March, Ouagadougou, 43 p.

17. Berton, S. (1987) Flood Control in the Lowlands. Small and Micro-Dams in West Africa [La maîtrise des crues dans les bas-fonds. Petits et micro-barrages en Afrique de l'Ouest]. Schedule 3rd-party, French Cooperation/GRET/ACCT, Paris, 302-474.

18. Couture, J.-L. (1999) Documentary Study on Water Management in Sub-Saharan and Caribbean Region [Etude documentaire sur la gestion de l'eau en Afrique sub-saharienne et région Caraïbes]. CTA-Wageningen, Pays-Bas, 41 p.

19. Global Water Partnership [Partenariat Mondial pour l'Eau] (2000) Integrated Water Resources Management. GWP/ Technical Advisory Committee [La gestion intégrée des ressources en eau. GWP/Comité Technique Consultatif (TAC)], Background Papers No. 4, Stockholm/ Sweden, 76 p.

20. Yabi, I., Afouda, F. and Boko, M. (2011) Some Socio-Economic Aspects of Irrigation Schemes in Response to Climate Change Impacts in Mono-Couffo Departments (Benin, West Africa) [Quelques aspects socio-économiques des aménagements hydro-agricoles en réponse aux impacts des changements climatiques dans les départements du Mono-Couffo (Bénin, Afrique de l'Ouest)], LACEEDE/DGAT/UAC, Cotonou, 13 p.

21. LIFAD (2006) Management Systems Study/Water Use and Definition of the Priority Actions of Local Water Resources Development in IWRM Approach, Benin. Volume 1-Current Situation of Water Resources Management in Benin [Etude des systèmes de gestion/utilisation de l'eau et définition des actions prioritaires de valorisation locale des ressources eau dans une approche GIRE au Bénin. Volume 1-Etat des lieux de la gestion des ressources en eau du Bénin]. MMEE/ LIFAD/Benin, 121 p.

22. MEE/DRH/HB (1998) RESO: Strategic and Prospective Diagnosis of Water Sector in the Southwest. Technical File, Project 7 [RESO: Diagnostic stratégique et prospectif du secteur de l'eau dans le Sud-ouest. Dossier technique, Projet 7]. ACP-BK.017, BF-UE, Ouagadougou, 52 p.

23. Diallo, M., Wellens, J., Dakoure, D. and Compaore, N.F. (2006) The Experience of Water Management Project in the Kou Basin in Burkina Faso: Promoting Innovative Approaches to Research and Development in Water [L'expérience du projet Gestion de l'Eau (GEau) dans le bassin du Kou au Burkina Faso: promouvoir des approches innovantes de recherche-développement en matière d'eau]. Unpublished paper, 11 p.

24. Carre, A., Peguin, M. and Poulin, B. (2012) Red List of IUCN Ecosystems. Application Exercise on Some Wetlands Ecosystems in Mainland France [Liste rouge des écosystèmes de l'UICN. Exercice d'application sur quelques écosystèmes de Zones Humides de France métropolitaine]. UICN/Tour du Valat, Paris/France, 82 p.

25. Mertz, O., D'haen, S., et al. (2012) Climate Variability and Environmental Stress in the Sudan-Sahel Zone of West Africa. AMBIO, Royal Swedish Academy of Sciences, 41, 380-392.

26. Dano Municipality [Commune de Dano] (2007) Municipal Development Plan. Final Edition [Plan Communal de Développement. Version définitive]. October, Dano, 128 p.

27. Pale, S. (2009) Farmers' Perceptions of Climate Variability, Their Impact on the Environment and Agricultural Activities in Dano Department (Southwestern Burkina Faso) [Perception paysanne des variabilités climatiques, leurs incidences sur l'environnement et les activités agricoles dans le département de Dano (Sud-ouest du Burkina Faso)]. Master Dissertation, Department of Geography, University of Ouagadougou, 96 p.

28. Roose, E., Albergel, J., De Noni, G., Laouina, A. and Sabir, M. (2008) Water Management Effectiveness and Soil Fertility in Semi-Arid Environments [Efficacité de la gestion de l'eau et de la fertilité des sols en milieux semi-arides]. AUF/IRD/ENFI/EAC, Paris, 425 p.

29. Salack, S., Sultan, B., Oettli, P., Muller, B., Gaye, A.T. and Hourdin, F. (2012) Rain Representation in Regional Climate Models and Application to Millet Yields Estimation in Senegal [Représentation de la pluie dans les modèles régionaux de climat et application à l'estimation des rendements du mil au Sénégal]. Sécheresse, 23, 14-23.

30. BUNASOLS (2000) Morpho-Pedological Study of Bougouriba and Ioba Provinces. Scale 1/1000.000—Technical Report No. 121 [Etude morphopédologique des provinces de la Bougouriba et du Ioba. Echelle 1/1000.000—Rapport technique N° 121], 69 p + annexes.

31. Dreyer Foundation and AGRITECH-Faso (2010). Integrated development program of the Municipality of Dano (PDIC/D) [Programme de développement intégré de la Commune de Dano (PDIC/D)]. Municipality of Dano/ AGRITECH-Faso/Dreyer Foundation, Dano, 117 p.

32. Roose, E. (2007) Restoring the Productivity of Tropical Soils [Restauration de la productivité des sols tropicaux]. Acts of JSIRAUF, Hanoi, 6-7 november, 6 p.

33. MAHRH (2008) Capitalization of Best Practices and Technologies in Irrigated Agriculture [Capitalisation des bonnes pratiques et technologies en agriculture irriguée]. SG/MAHRH—ARSA/PNUD, june, Ouagadougou, 168 p.

34. MAHRH (2004) Legal Status of River Basin Organizations in Burkina Faso and Institution of the Financial Contribution in Terms of Water According to the Orientation Law Related to Water Management [Statut juridique des organismes de bassins hydrographiques du Burkina Faso et Institution de la contribution financière en matière d'eau selon la loi d'orientation relative à la gestion de l'eau]. SG/DGIRH, nov. Ouagadougou, 61 p.

35. Some, J.-M. (2013) My Fight for Land [Mon combat pour la terre]. KARTHALA Editions, Paris, 256 p.

36. Ballo, Y., Coulibaly, A., Alassane, D. and Yattara, K. (2006) Issues of Equality Lowland Resources Access. Evidence, Importance, Factors and Impacts on Lowlands Development [Problématique d'accès des genres aux ressources des bas-fonds. Témoignages, importance, facteurs et impacts sur l'aménagement des bas-fonds]. DIC-PT N°26, DDC/ Intercooperation Bern et DIC au Sahel, Ouagadougou, 22 p.

37. Kuba, R. (2004) The Ritual Grammar Hierarchies: Migration and Land Chiefs in Segmental Society (Burkina Faso) [La grammaire rituelle des hiérarchies: migrations et chefs de terre dans une société segmentaire (Burkina Faso)]. Autrepart, No. 30, 63-76.

38. Some, M. (1999) The Sacred and the Traditional Political System of Dagara in Burkina, on Colonization Toil [Le sacré et le système politique traditionnel des Dagara du Burkina à l'épreuve de la colonisation]. Revue CAMES: Sciences sociales et humaines—Serie B, 1, 74-89.

39. CODDE (2006) The Release and Subsidies Impact on Rice Sector in Burkina Faso [L'impact de la libération et des subventions sur la filière riz au Burkina Faso]. Ouagadougou, 95 p.

40. MEE/GIRE (2000) Sociological Issues Concerning Water Resources Management in Burkina Faso. Sociological Technical Report No. 2

[Problématiques sociologiques concernant la gestion des ressources en eau du Burkina Faso. Rapport technique sociologique N°2]. MEE/IWRM Program, December, Ouagadougou, 163 p.

41. Fromageot, A. (2007) Vegetable Colonization of Small Dams Shores: A New Geography [Colonisation maraîchère des rives des petits barrages: une nouvelle géographie]. 098HYSOC, L'eau en partage, IRD/Abidjan, 229-243.

42. Petitguyot, T., Rieu, T., Chohin-Kuper, A. and Doukkali, R. (2005) Irrigated Agriculture Modernization and Water Resources Sustainability in the Scope of Tadla in Morocco [Modernisation de l'agriculture irriguée et durabilité des ressources en eau dans le périmètre de Tadla au Maroc]. Acts of Seminar "Irrigated Agriculture Modernization", Rabat, 13 pages.

43. Saidi, A., Hammami, M., Daghari, H., Hatira, A., Ltifi, W. and Hachani, K. (2010) Water and Salts Balances in the Irrigated Perimeter of Kalaat Landalous (Tunisia) [Bilans d'eau et des sels dans le périmètre irrigué de Kalaat Landalous (Tunisie)]. Sécheresse, 21, 1-5.

44. Traore, B., Ouedraogo, O., Miwende, S. and Sanou, M. (2008) Diagnostic Study of Moutori Developed Plain: Farmers' Capacity Development Plan [Etude diagnostic de la plaine aménagé de Moutori: Plan de développement des capacités des producteurs]. Dreyer Foundation /Municipality of Dano, Dano, 75 p + Annexes.

45. DGPER (2011) National Strategy for Rice Development [Stratégie nationale de développement de la riziculture]. Ouagadougou/Burkina Faso, 43 p.

46. Romagny, B. and Cudennec, C. (2006) Water Management in Arid Areas: Physical and Social Considerations to Identify the Relevant Territories in South-East Tunisia [Gestion de l'eau territoires pertinents dans le en milieu aride: Considérations physiques et sociales pour l'identification des Sud-Est tunisien]. http://developpementdurable.revues.org/1805

47. Zammouri, A. and Dababi, M.S. (2007) Wild Runoff Control in Arid Areas by Conventional Retaining Structures in the Troughs (Jessours) and Storage Tanks (Majels and Fasguyas) [La maîtrise du ruissellement sauvage en zones arides par des ouvrages de retenue traditionnels dans les Talwegs (jessours) et des Citernes de Stockage (Majels et Fasguyas)]. ADD, Tunisie, 7 p.

Chapter 11

Water Allocation Systems

Ralph A. Wurbs[1]

[1] Texas A&M University, USA

1. INTRODUCTION

Water resources are shared by many people who use the water for a variety of purposes. Water allocation systems serve to equitably apportion water resources among users; protect existing water users from having their supplies diminished by new users; govern the sharing of limited water during droughts when supplies are inadequate to meet all needs; and facilitate efficient water use. Effective water allocation becomes particularly important as demands exceed reliable supplies. As water demands increase with population and economic growth, water allocation systems must be expanded and refined.

The institutional framework for water resources development and management involves a hierarchy of water allocation systems. Water resources are allocated between nations by treaties and other agreements. In the United States (U.S.), water is allocated between states through interstate compacts. Water is shared by regional water authorities, municipal utility districts, cities, private companies, irrigation districts, farmers, and individual domestic water users through water rights systems. Water supply entities service their customers in accordance with contracts and other commitments. Federal water development agencies provide reservoir storage capacity for nonfederal sponsors.

This chapter begins with a general overview of institutional systems for allocating the water flowing and/or stored in rivers, lakes, and aquifers to diverse types of water use and numerous water users. The chapter then focuses on the Texas experience in implementing water allocation systems, including both accomplishments and issues still remaining to be resolved. The state of Texas in the United States serves as a case study that illustrates concepts and strategies that are relevant throughout the world.

With a population of 26 million people and land area of 696,000 km^2, Texas is a large state with diverse geography, economy, climate, hydrology, and water management practices. Texas has a rich heritage of implementing water allocation strategies as a central thrust of its water resources planning, development, and management. The Rio Grande is shared with Mexico, and several major river basins are shared with neighboring states in the U.S. Thousands of government agencies, cities, private companies, and citizens hold rights to use the waters of the state. Water allocation involves multiple overlapping institutional mechanisms. The state has progressed significantly in recent years in improving its water allocation systems. Further refinements continue to be a priority policy emphasis.

The Water Rights Analysis Package (WRAP) modeling system simulates the water allocation systems described in this chapter. WRAP is routinely applied in Texas in regional and statewide planning studies, administration of the water rights permit system, and other water management activities. WRAP is generalized for application anyplace in the world.

2. WATER ALLOCATION MECHANISMS

People in various nations, regions, and local communities have developed their own sets of institutions and practices governing the sharing of water. These water allocation systems have evolved historically and continue to change. Hierarchies of water allocation systems in the U.S. and many other countries generally have the following components or features.

- The waters of international rivers and aquifers are allocated between nations based on international law, customs, treaties, and agreements.

- In the U.S., waters of interstate river basins are allocated between states based on compacts negotiated by the states and approved by the federal government.

- Certain rights are reserved for military installations and other government owned lands and facilities.

- A legally established priority system based generally on variations of the riparian or prior appropriation doctrines guides the allocation of the surface water flowing in streams and stored in reservoirs.

- A separate system of laws and customs guides the allocation of water resources in ground water aquifers.

- An administrative system that grants, limits, and modifies water rights and enforces the allocation of water resources may or may not include formal issuance of written permits to water right holders.

- Water users and water management entities implement various contracts and other formal agreements.

- Sharing of water resources may be governed by cultural traditions and informal agreements that evolve historically over many years.

Water allocation mechanisms typically vary greatly between ground-water and surface-water. From a water law perspective, ground and surface water are usually treated as separate resources. The extent to which the important hydrologic and water management interconnections are recognized varies between geographical regions.

The institutional mechanisms of water allocation are typically viewed from policy, legal, economic, and social perspectives. However, hydrologic science and engineering are also important aspects of developing and maintaining water allocation systems [1, 2, 3].

2.1. Allocation Of The Waters Of International Rivers And Aquifers

Principles and rules of international water law are found in treaties, international custom, general principals of law, and writings of international institutions [4, 5, 6]. Two hundred and sixty-one international river basins, each encompassing portions of two or more nations, cover about 45 percent of the world's land area excluding Antarctica [7]. Little progress has been made in developing effective water allocation systems in many of these international river basins. Water allocation is even more difficult for groundwater aquifers shared by two or more countries.

Effective joint multiple-nation water management will be a major determinant in achieving stability, peace, and prosperity in many regions of the world in the 21st century [8, 9, 10, 11, 12]. Examples of the many regions with dramatic potential for either cooperation or conflict include the following. The Jordan River shared by Israel, Jordan, Syria, and the Palestinians is a small stream with remarkably great historical and political importance. Shared groundwater is also an important issue this region. The Euphrates and Tigris Rivers flow from Turkey through Syria, Iraq, and Iran. Most of the flow of the Euphrates and Tigris Rivers originate in their upper watersheds in Turkey and is controlled by a system of major dams in Turkey. The Ganges and Brahmaputra River Basins in Nepal, China, India, Bhutan, and Bangladesh, with a history of centuries of water conflicts, contained an estimated 400 million people in 2000 living at an impoverished standard of living. In the Southern African Region encompassing Angola, Botswana, Lesotho, Malawi, Mozambique, Namibia, South Africa, Swaziland, Tanzania, Zambia, and Zimbabwe, every major river is shared by two or more nations. Population growth and economic development are resulting in intensified demands on limited water resources with a long history of controversy.

2.2. Water Rights In The United States

A water right is the legal privilege to store, regulate, and/or divert water for beneficial use. Water law is the creation, allocation, and administration of water rights. Institutions and customs for executing water rights play key roles in water management in various regions throughout the world. The following discussion is from a U.S. perspective but has broader applicability in other countries as well.

Books on water rights/law range from a concise book entitled *Water Law in a Nutshell* [13] to a massive eight volume collection [14] first published in 1967 and again as a revised updated edition in 1991 with the individual volumes continuing to be periodically updated and expanded at different times. The American Society of Civil Engineers has published guidelines for developing water right rules and regulations [15, 16, 17]. Other books focus on water rights in particular regions of the world other than the U.S. [18, 19, 20, 21].

Each state in the U.S. has developed its own set of laws, institutions, and practices governing water rights. These systems have evolved historically and continue to change. States in the western and eastern halves of the U.S. have generally adopted different approaches to water rights due largely to the western states having much drier climates. Water allocation and accounting systems tend to be more rigorous in regions where demands approach or exceed supplies. The experience of the state of Texas discussed in this chapter illustrates key aspects of developing and administering water right systems. Regions of Texas are representative of both western and eastern states.

Surface water in streams and lakes is viewed in the U.S. as a renewable resource owned by the state and used by the public. Groundwater is typically viewed as the property of the owners of the land overlying the aquifers. Water rights for groundwater aquifers are very different than for surface water streams and lakes. Although water rights are established primarily at the state level, federal laws govern water rights for military installations, other federal lands such as national parks, and Indian reservations.

For river basins encompassing portions of multiple states, water is allocated between states based on interstate compacts developed by the states, approved by the U.S. Congress, and administered by interstate compact commissions. Disputes may arise in implementing interstate compacts, particularly during droughts. Disputes are settled by the U.S. Supreme Court if compact commissions cannot work out disagreements between the states through negotiation. A major objective of compact commissions is to avoid lawsuits, but disputes in the western states in particular often reach the litigation stage. Various states have experienced lengthy and costly water allocation disputes settled through the U.S. Supreme Court [22]. Shared management of interstate groundwater aquifers and associated development of water allocation mechanisms have not progressed to nearly the extent as surface water.

2.2.1. Surface Water Rights

Legal rights to the use of stream flow are generally based on two alternative doctrines, riparian and prior appropriation. The basic concept of the riparian doctrine is that water rights are incidental to the ownership of land adjacent to a stream. The prior appropriation doctrine is based on the concept of protecting senior water users from having their supplies diminished by newcomers developing water supplies later in time. In a prior appropriation system, water rights are not inherent in land ownership, and priorities are established based on dates that water is appropriated.

The doctrine of riparian rights common in the eastern U.S. is based on English common law. Under the strictest interpretation of the riparian doctrine, the owner of riparian land adjacent to a stream is entitled to receive the full natural flow of the stream without change in quantity or quality. Since a strict interpretation imposes impractical constraints on water use, the riparian doctrine is normally interpreted to allow riparian land owners to divert reasonable amounts of stream flow for beneficial purposes. Varying definitions of *reasonable amounts* complicate water allocation.

The doctrine of prior appropriation is associated with settling the American West. As settlers moved from the eastern states to the West in the 1800's, farmers and ranchers claimed land, and miners claimed gold and other minerals. Likewise, water was appropriated by the first to arrive and claim the resources for beneficial use. In developing their farms and communities, people needed protection from having their water supplies diminished with later population growth and economic development.

Most of the western states have established permit systems in which a state agency issues permits to water right holders specifying amounts and conditions of water use. Riparian and/or appropriative rights may be incorporated into the original development of the permit system, with additional new permits being issued based on prior appropriation. With growing demands on limited water resources, permit systems will likely continue to be developed in the eastern states, which have more abundant stream flow, similar to those already in place in the drier western states.

Surface water rights in the eight driest western states (Nevada, Arizona, Utah, Idaho, Montana, Wyoming, Colorado, and New Mexico) are based purely on the prior appropriation doctrine. Alaska is also a prior appropriation state though somewhat different. Ten western states with hybrid systems merging riparian and appropriative rights into permit systems include California, Oregon, Washington, Texas, Oklahoma, Kansas, Nebraska, South Dakota, North Dakota, and Mississippi. Hawaii has a unique hybrid system. Water rights in 30 eastern states are based primarily on the riparian doctrine [23].

2.2.2. Groundwater Rights

The rights and obligations for groundwater use are generally tied to two legal principles: property ownership and shared ownership of a common public resource. A variety of state approaches to groundwater rights has evolved from these concepts. State groundwater law is based on mixtures of the following doctrines.

Absolute Ownership Doctrine: Landowners own the groundwater under their land and may drill wells and pump as much water as they wish. Texas and several other states have historically adhered to this doctrine but are slowly changing.

Reasonable Use Doctrine: Landowners own groundwater, but their pumping is limited to reasonable use which has been defined in a variety of ways. This doctrine is common in the eastern states.

Correlative Rights Doctrine: In times of shortage, groundwater is shared by overlying landowners in proportion to the amount of land they own. This extension of the reasonable use rule is primarily associated with California.

Prior Appropriation Doctrine: Groundwater is allocated similarly to surface water with priorities assigned based on the dates that users first appropriate the water for beneficial use. This doctrine is common in the western states.

Permit Systems: Systems in which state agencies issue permits specifying the amounts and conditions of water use have been adopted in 33 of the 50 states. The other doctrines may be reflected in the water rights documented by the permits. Texas is among the 17 states that have no state-wide groundwater permit program.

Some states divide groundwater into categories with different water right rules applied to each classification. Percolating groundwater may be legally differentiated from underground streams with definable flow paths. Underground streams are sometimes treated as being similar to surface streams.

The issue of impacts of groundwater pumping on surface stream flow has been addressed to varying extents in different states. In some states, groundwater is classified as either tributary or nontributary. Tributary groundwater hydrologically contributes to surface stream flow. Nontributary groundwater does not. Water right rules and management strategies for tributary groundwater are based on protecting surface water rights.

3. WATER ALLOCATION SYSTEMS IN TEXAS

Texas is a large state located in the south-central U.S. that is representative of both the drier western and wetter eastern states from various perspectives including climate. Mean annual precipitation varies from 20 cm at the city of El Paso in arid west Texas to over 140 cm in the humid eastern extreme of the state. Texas actually provides two case studies since water allocation in the Lower Rio Grande Valley has distinct differences from the remainder of the state. Water rights are a major consideration in river basin management statewide. Allocation of ground water is very different than allocation of surface water.

Water resources development and management in Texas is governed largely by the need to be prepared for extended droughts. The hydrologically most severe drought of record began gradually in 1951 and ended in 1957 with one of the largest floods on record. Droughts in the 1910's and 1930's were also extended multiple-year dry periods over large areas of Texas and neighboring states. The drought that occurred during 1995-1996 was much more economically costly than the earlier droughts due to the population and economic growth that had occurred. In terms of annual precipitation, for more than half of the land area of Texas, 2011 was the driest calendar year since the beginning of official observed precipitation records in 1895. The remainder of the state was also very dry during 2011. Severe drought conditions are continuing during 2012 throughout Texas and other states in the U.S.

3.1. Water Resources Planning, Development, and Management

The Texas Water Development Board (TWDB) in partnership with 16 regional planning groups outline future water resources needs and challenges in the 2012 State Water Plan, which is presented in a multiple volume report entitled *Water for Texas 2012* [24]. Water demands are projected to increase about 22 percent between 2010 and 2060. Available water supplies with existing infrastructure and current institutional arrangements will decrease about 10 percent during this period due to reservoir sedimentation and depletion of groundwater aquifers. Environmental flow requirements and ecosystem preservation are major concerns as well as meeting municipal, industrial, and agricultural water needs. The regional planning groups have identified several hundred water supply projects with an estimated cost of over \$50 billion to address intensifying water needs. The TWDB predicts that annual losses from not meeting water supply needs could result in a reduction in income of approximately \$12 billion annually if current drought conditions approach the drought of record, and as much as \$116 billion annually by 2060 [24].

The map of Texas in Figure 1 shows the larger rivers and cities of the state. The state encompasses a land area of 696,000 km^2 divided into 15 major river basins and eight coastal basins located along the Gulf of Mexico between the lower reaches of the major river basins. The 1990 population of 17.0 million people increased to 25.7 million in 2011 and is projected by the Texas Water Development Board [24] to increase to 46.3 million by 2060. Fifty-eight percent of Texans live in the state's three largest metropolitan areas: the Dallas/Fort Worth Metroplex in the upper Trinity River Basin, greater Houston area in the San Jacinto River Basin, and city of San Antonio in the San Antonio River Basin. With a 2011 population of 6.5 million people, the Dallas/Fort Worth (DFW) Metroplex is the largest metropolitan area in the southern U.S. and the fastest growing metropolitan area in the U.S. With a 2011 population of 6.1 million, the city Houston and adjacent smaller cities is also one of the largest and fastest growing metropolitan areas in the U.S. San Antonio is the third largest city in Texas with a population of 2.2 million people. Conversely, several of the major river basins of the state have extremely low population densities. Water management practices are very diverse throughout the state.

Total diversions of 22 billion m^3/year from streams, reservoirs, and aquifers statewide in 2010 was for agricultural irrigation (56%), municipal use (27%), industrial use (15%), and livestock (2%). Currently, about half of the water supplied is from groundwater aquifers, and the other half is from surface streams and reservoirs. However, problems caused by decades of groundwater pumping rates significantly exceeding recharge rates in various regions of the state are resulting in major shifts toward greater reliance on surface water.

The naturalized flow volumes during each of the 864 months of 1940-2011 at a gauging station on the Brazos River near Houston are plotted in Fig. 2 to illustrate the tremendous variability that characterizes stream flows throughout Texas. A highly variable resource is allocated to numerous water users. Long-term mean flows may be relatively large, but most of the flow occurs during infrequent flood events. The water management community must deal with droughts with durations ranging from several months to several years. Dams with large storage volumes are essential in order to develop dependable water supplies. The river flows plotted in Fig. 2 are generated by the modeling system described later in this chapter by adjusting gauged flows to represent natural conditions without human water resources development and use.

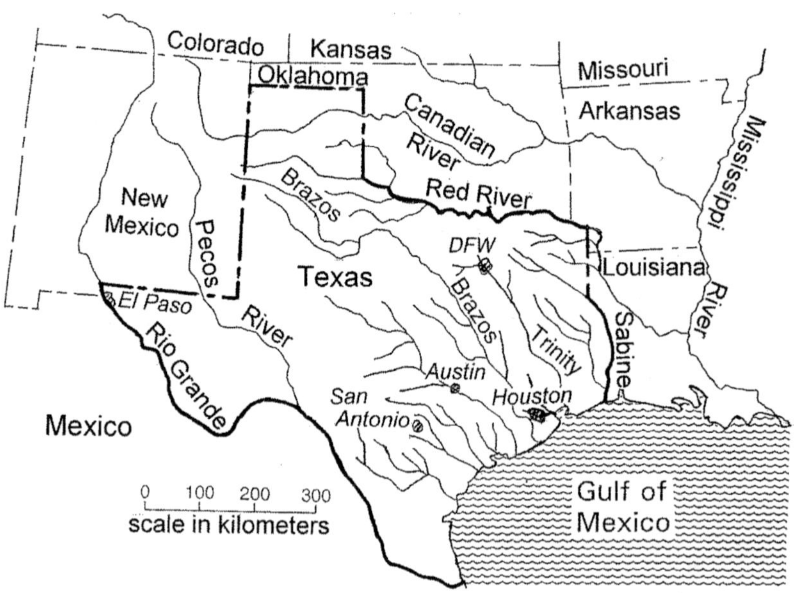

Figure 1. Map of Texas showing major rivers, largest cities, and neighboring states

Conservation and flood control storage capacities totaling 50 and 23 billion m^3 are provided by 211 major reservoirs with individual capacities of 6.2 million m^3 or greater. Although there are several thousand smaller reservoirs, the 211 major reservoirs account for over 95% of the storage capacity in the state. Texoma on the Red River and International Amistad on the Rio Grande are the largest reservoirs with capacities of 6.6 and 6.3 billion m^3. The 22 hydroelectric plants in the state are components of electric utility systems dominated by thermal plants. Reservoir releases through hydropower turbines are almost always incidental to downstream water supply needs.

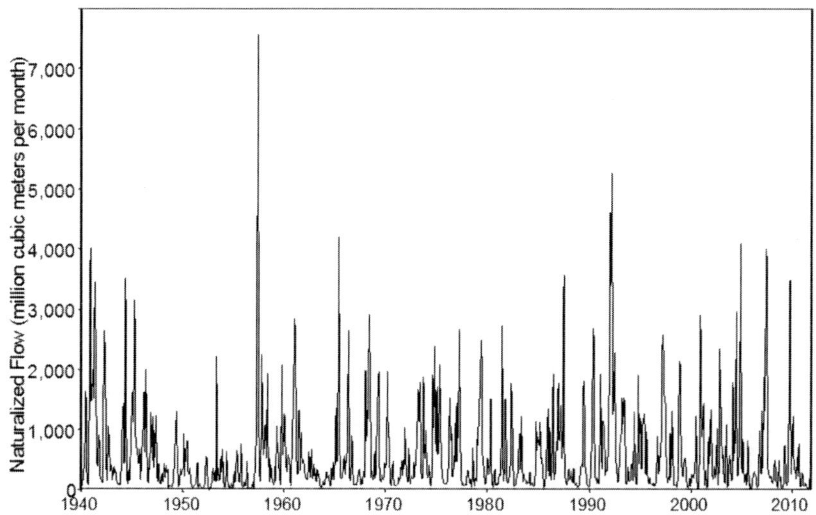

Figure 2. Monthly naturalized flows of the Brazos River illustrating variability

Federal involvement in developing the state's water resources includes 32 U.S. Army Corps of Engineers (USACE) reservoirs in eight river basins and two International Boundary and Water Commission (IBWC) reservoirs on the Rio Grande that account for about 40% and 90% of the conservation and flood control, respectively, storage capacity of the 211 major reservoirs. The U.S. Bureau of Reclamation constructed three reservoirs that are now owned by local entities. River authorities and cities have contracted for the water supply storage capacity of the USACE reservoirs. All costs allocated to water supply are reimbursed by the nonfederal sponsors, who also hold the water right permits. The USACE is responsible for flood control operations.

Nineteen river authorities are responsible for development and management of the water resources of all or portions of major river basins. River authorities construct and operate their own reservoir projects and contract for storage capacity in federal reservoirs. Numerous municipal utility districts and irrigation districts also play key roles in supplying water from rivers and reservoirs. Groundwater conservation districts facilitate groundwater management.

Some type of water rights system has been administered statewide by a centralized agency since 1913, but that agency has changed over time. The Board of Water Engineers was established in 1913; reorganized as the Texas Water Commission (TWC) in 1962; and renamed the Texas Water Rights Commission in 1965 with non-water rights functions being transferred to the Texas Water Development Board (TWDB), which had been previously created in 1957. In 1977, the Texas Department of Water Resources (TDWR) was created by combining the Water Rights Commission, TWDB, and Water Quality Board. In 1985, the TDWR was dissolved, and the TWC and TWDB became separate agencies. The Texas Natural Resource Conservation Commission (TNRCC) was created in 1993 by merging the TWC and Texas Air Quality Board. The TNRCC was renamed the Texas Commission on Environmental Quality (TCEQ) in September 2002. The TCEQ is now responsible for administering the many regulatory programs of the state including the water rights system.

The TCEQ consists of three full-time commissioners appointed by the governor and a professional and administrative staff of over 3,000 employees. Water rights are one of many regulatory responsibilities of the TCEQ. The TCEQ and TWDB interact closely in many of their activities. The TWDB is responsible for developing and updating the State Water Plan and administering an array of financial assistance programs. The TWDB has a governing board of six members appointed by the governor and a staff of about 800 employees.

3.2. Legislatively Mandated Water Management Programs

The state-level water allocation systems described in this chapter were created under laws enacted by the Texas Legislature. As discussed later, the statewide water rights permit system for all of Texas except the Rio Grande was implemented pursuant to the Water Rights Adjudication Act of 1967. The water rights adjudication process mandated by the 1967 Act was completed shortly before a severe statewide drought in 1995-1996 which motivated enactment of the 1997 Senate Bill 1.

The Brown-Lewis Water Management Plan enacted by the Texas Legislature as its 1997 Senate Bill 1 is considered to be a milestone in the history of water resources management in Texas. The designation*Senate Bill 1* is traditionally applied each legislative session to highlight legislation of the upmost importance. The 1997 Senate Bill 1 and subsequent amending legislation authorized the Water Availability Modeling (WAM) System described later in this chapter, a regional and statewide water resources planning process noted below, and other related water management activities.

The 1997 Senate Bill 1 established a program for developing regional water plans that are integrated into a statewide planning process administered by the Texas Water Development Board (TWDB). Committees of local water interests have been established to prepare plans for the orderly development, management, and conservation of the water resources of each of 16 regions. The TWDB provides funding, administrative, and technical support to the regional committees. Consulting firms perform much of the technical work. The 1997 Senate Bill 1 mandated that initial regional plans be completed and incorporated into a statewide plan by 2002. Continuing planning is organized based on updated plans being reported at cycles of not to exceed five years. The 2002, 2007, and 2012 state water plans consists of 16 regional reports and a statewide report which are publically available at the TWDB website.

In evaluating water right permit applications, the TCEQ requires that proposed actions must be consistent with pertinent regional plans. The WAM System is routinely applied in both regional and statewide planning and administration of the water right permit system.

The Texas Instream Flow Program [25] established by Senate Bill 2 (SB-2) enacted by the Texas Legislature in 2001 and expanded by Senate Bill 3 (SB-3) in 2007 is administered jointly by the TCEQ, TWDB, and Texas Parks and Wildlife Department (TPWD). The purpose of the SB-2 program is to perform scientific studies to determine flow conditions necessary for supporting a sound ecological environment in the river basins of Texas. SB-3 mandated a new regulatory approach for protecting environmental flows through a local stakeholder process culminating in rules to be administered through the TCEQ. The objectives of the 2001 SB-2 and 2007 SB-3 are being accomplished largely through the ongoing work of Bay and Basin Expert Science Teams (BBEST) and Bay and Basin Area Stakeholder Committees (BBASC) for the individual river basins, representing the scientific and water management/use communities, with technical support from consultants.

The TCEQ, TWDB, and TPWD provides administration oversight and technical support.

Environmental instream flow requirements are determined by the BBEST teams for their assigned streams within a framework of subsistence flows, base flows, high flow pulses, and overbank flood events needed to support riverine ecosystems, wetlands, and freshwater inflows to bays and estuaries. The Stakeholder Committees consider the BBEST recommendations within the content of municipal, industrial, and agricultural water needs as well as environmental water needs. The TCEQ is responsible for final approval and incorporation of the recommendations of these groups and the public into the water rights permit system. As of 2012, the endeavors of the Texas water management community, under the leadership of the three-agency partnership and the BBEST and BBASC groups, to better incorporate environmental considerations into the water rights permit system is well underway and is expected to continue for many years into the future.

3.3. Groundwater Management

Most of Texas is underlain by 9 major and 21 minor aquifers that currently supply about half of the water used in the state. Depleting ground water reserves due to excessive pumping is a major problem throughout the state and is forcing a shift from groundwater use to a greater reliance on surface water supplies. Groundwater rights in Texas have historically been based on the common law rule allowing land owners to pump as much water as they wish from under their land. However, increased state regulation of groundwater is evolving over time primarily through the establishment of groundwater conservation districts.

In 1949 and 1985, the Texas Legislature passed laws authorizing creation of groundwater conservation districts with local voter approval. The 1949 legislation allows local residents to petition the state to have a district created. The 1985 amendment authorizes the TCEQ to designate and initiate formation of districts for areas with critical problems. Local voters can still veto a proposed district, but if they do, state funding for water projects can be withheld. Twelve districts existed prior to 1985. As of 2012, 96 groundwater districts covering over half of the land area of the state are operational. Three additional districts are currently in the process of being confirmed by voters through local elections at the county level. A total of

174 of the 254 counties in Texas are within a groundwater conservation district.

The primary purposes of the districts are to encourage water conservation and to protect water quality. Most districts direct their efforts toward prevention of waste, water conservation education, recharge projects, and data collection. Some are now moving toward stricter regulation of groundwater use. The districts tread a narrow path between private ownership of groundwater and state responsibility to protect the water resource. Texans are reluctant to allow anyone to tell them how much water they can pump from under their own land. Governmental regulation of pumping has been driven by necessity as depleting aquifers resulted in major problems. The Harris-Galveston Coastal Subsidence District and Edwards Underground Water Authority have developed the strongest regulatory programs.

The Harris-Galveston Coastal Subsidence District was created in 1957 in response to severe subsidence in the vicinity of the cities of Houston and Galveston. Due to decades of over-drafting groundwater, the ground surface has been lowered over three meters in places in this low-lying, heavily urbanized coastal region. Houston and neighboring smaller cities continue to shift from ground water to surface water supplies. Groundwater pumping is regulated by the Subsidence District through a permit program.

The Edwards Aquifer Authority was created in 1993 largely due to a federal court ruling related to protection of endangered species under the Endangered Species Act. The Edwards is a limestone aquifer shared by San Antonio, several smaller cities, and extensive irrigated farming interests. San Antonio is the largest city in the U.S. that relies solely on groundwater for its water supply. Springs fed by the Edwards Aquifer maintain the flow of several rivers and support ecosystems that include several endangered species. The Edwards Aquifer Authority administers a water rights permitting system to facilitate aquifer management. Pumping limits are reduced whenever water levels in selected wells fall below specified criteria.

All of the groundwater conservation districts are required to develop, implement, and periodically update management plans for the effective management of their groundwater resources. These plans are subject to approval by the TWDB and are publically accessible through the TWDB website.

Although interstate and international aquifers are important, allocation agreements do not exist. The U.S. and Mexico have different legal reasons for their lack of management of ground water [26]. The Mexican federal government has the authority to enforce comprehensive management and regulation of ground water but has not chosen to do so. In the U.S., the legal framework is inadequate and chaotic. Another constraint to joint U.S. and Mexico management of ground water is the lack of clarity of international ground water law [26].

3.4. Allocation Of The Waters Of The Rio Grande

The Rio Grande (Spanish for Large River) is unique relative to the other river basins of Texas from several perspectives. It is shared by two nations. The Lower Rio Grande Valley accounts for the majority of the surface water irrigation in Texas. The intensive agricultural production of the region depends almost exclusively on the Rio Grande with little use of groundwater. Other major irrigation regions of the state rely primarily on groundwater. The water rights system for the Lower Rio Grande was developed separately and has distinct differences from the remainder of the state, particularly in regard to the priority system and water master operations. Fort Quitman shown in Fig. 3, located 140 km downstream of the city of El Paso, is used as the divide between the lower and upper portions of the basin in both the international and state water allocation systems.

The Rio Grande Basin is shared by Mexico and three states in the U.S. Water allocation is governed by two international treaties and two interstate compacts. Allocation of the Texas share of the waters of the Rio Grande to irrigators, cities, and other users is based on a water rights permit system governed by state law. The Rio Grande Compact approved by the legislatures of Colorado, New Mexico, and Texas in 1939 allocates the uncommitted waters of the Rio Grande above Fort Quitman. The Pecos River Compact adopted in 1948 allocates the waters of that tributary between Texas and New Mexico.

3.4.1. Mexico-United States Treaties

A 1906 treaty between the U.S. and Mexico provides for delivery of 74 million m^3/yr of Rio Grande water to Mexico in the El Paso-Juarez Valley above Fort Quitman. Elephant Butte Reservoir in New Mexico, operated by the Bureau of Reclamation, and the American and International diversion dams near El Paso, operated by the IBWC are used to implement the water allocation provisions of the treaty. The treaty further provides that if water is unavailable, the amount allocated to Mexico shall be diminished in the same proportion as water delivered to irrigate lands in the United States. This provision has been invoked in about a third of the years since 1951.

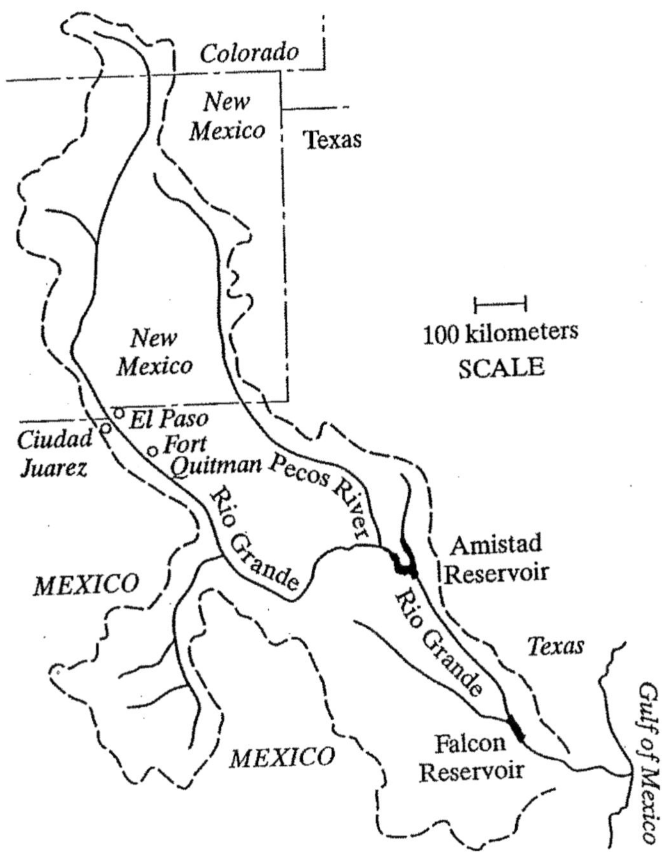

Figure 3. Rio Grande Basin

The Water Treaty of 1944 expanded the International Boundary Commission to the International Boundary and Water Commission (IBWC), provided for the distribution of waters of the Rio Grande from Fort Quitman to the Gulf of Mexico between the two nations, and authorized construction of International Amistad and Falcon Reservoirs.

The 1944 treaty administered by the IBWC also includes provisions for allocation of the waters of the Colorado River. The Colorado River drains 629,000 km^2 in seven western states and flows into the Gulf of California in Mexico. Although both are included in the same 1944 treaty, water allocation for the Colorado River and Rio Grande are very different. The Rio Grande serves as a boundary between the two countries. The Colorado River flows from the U.S. to Mexico with most of its watershed area being located in the U.S.

The IBWC is composed of a Mexico Section with commissioner and technical and administrative staff located in Ciudad Juarez and a U.S. Section with offices across the river in El Paso. Amistad and Falcon Reservoirs are operated by the IBWC primarily for flood control and water supply for the Lower Rio Grande Valley. Hydroelectric power generation uses releases for downstream water supply. The U.S. share of the conservation storage is used to meet demands in the lower basin in accordance with the state water rights system.

Conservation and flood control capacities are 4.17 billion m^3 and 2.15 billion m^3 in Amistad Reservoir and 3.29 and 0.63 billion m^3 Falcon Reservoir. In accordance with the 1944 treaty, the U.S. has 56.2% and 58.6% of the conservation storage capacity of Amistad and Falcon, respectively, with Mexico owning the remainder. The IBWC operates Anzaldus and Retamal Dams on the lower Rio Grande to facilitate diversions. The travel time for releases from Falcon Reservoir to reach the most downstream diversion locations is about a week.

The IBWC maintains a continuous accounting of the volume of stored water owned by each of the two countries. Stream flows into Falcon and Amistad Reservoirs are allocated between the two countries. Flows on a number of major tributaries named in the treaty are gauged and allocated as specified by the treaty. All other flows not otherwise allocated are divided equally between the two countries. Mexico receives all the flows from several specified Mexican tributaries and two-thirds of the flows from other specified tributaries. The U.S. receives all the flows from certain U.S. tributaries and one-third of the flows from the other specified tributaries.

Computations are performed routinely to allocate the reservoir inflow and evaporation volumes, which are combined with recorded releases to determine the amount of water that each country has in storage.

The 1944 treaty gives the U.S. a right to one-third of the flow reaching the Rio Grande from six Mexican tributaries, provided that this third shall not be less, as an average amount in cycles of five consecutive years, than 431,721,000 m^3 annually. Other provisions relate to special conditions. A significant deficit accumulated over several years during the 1900's-2000's in meeting these requirements. Discussions addressing the stream flow deficit owed by Mexico to the U.S. included requests by the U.S. that operating policies for reservoirs in Mexico be modified to mitigate the deficit.

3.4.2. Allocation Of The Texas Share Of The Lower Rio Grande

The Texas share of the waters of the Rio Grande below Fort Quitman was allocated among numerous water rights holders in conjunction with a massive lawsuit, commonly called the Lower Rio Grande Valley Water Case. The lawsuit was filed in 1956, the trial was held in 1964-1966, and the final judgment of the appellate court was filed in 1969. In 1971, the Texas Water Rights Commission adopted rules and regulations implementing the court decision. Assorted versions of riparian and appropriative rights were combined into a permit system. The litigants in the Rio Grande law suit included 42 water districts and 2,500 individuals. More than 90 lawyers appeared before the court. The expense and effort demonstrated the impracticality of a purely judicial determination of water rights for the entire state and led to enactment of the Water Rights Adjudication Act of 1967.

The lawsuit resulted in water rights being divided into three categories. Municipal rights have the highest priority. Irrigation rights are divided into Class A and Class B rights, with Class A rights receiving more storage in Falcon and Amistad Reservoirs storage accounts in the allocation procedure. Although this weighted priority system for irrigation rights has little significance during years of plentiful water, its effect in water-short years is to distribute the shortage among all users, with the greater shortages occurring on lands with Class B water rights.

Most of the U.S. share of the water regulated by Amistad and Falcon Reservoirs is used in the very productive agricultural region of Texas below Falcon Reservoir. Irrigation districts, individual farmers, and cities communicate their water needs to the TCEQ Rio Grande Water Master Office, which in turn schedules releases from Falcon and Amistad Reservoirs with the IBWC. The Water Master Office maintains an accounting of the amount of water used and the amount of water in reservoir storage allocated to each of about 1,600 water rights accounts.

The allocation rules administered by the TCEQ Water Master first provide a reserve of 278 million m^3in Amistad and Falcon Reservoirs for domestic, municipal, and industrial uses. Next, available water is allocated to an operating reserve that provides for seepage and evaporation losses, adjustments required as the IBWC computations of Mexico-U.S allocations are updated periodically, and emergency requirements. The remaining water in storage is allocated among all the irrigation permit holders. The storage is basically allocated in proportion to annual diversion rights, except the Class A rights are multiplied by a factor of 1.7 to allow them a greater storage allocation than Class B rights. Other provisions include limiting each storage allotment to not exceed more than 1.41 times its authorized diversion right. If an irrigation right does not use water for two consecutive years, its storage account is reduced to zero.

3.5. Interstate River Compacts

Texas participates in five interstate river compacts executed by the member states and approved by the U.S. Congress. The rivers and the dates the compacts became effective are Rio Grande, 1939, Pecos, 1948, Canadian, 1952, Sabine, 1954, and Red, 1980. The purposes of the compacts are to provide for equitable allocation of water between the states and to facilitate cooperative planning and management. Commissions with a representative from each member state administer the compacts. The commissioners have minimal staffs and rely on the state water agencies for technical support.

Texas filed a lawsuit against New Mexico in 1974, claiming that Texas did not receive its full share of water allocated by the 1948 Pecos River Compact due to over-use in New Mexico [27]. In 1988, based on the recommendations of a court appointed Special Master, the Supreme Court awarded Texas a cash settlement for past damages and established a Pecos River Master to administer future water allocations under the compact.

3.6. Statewide Surface Water Rights Permit System

All of Texas except the Lower Rio Grande below Fort Quitman has a consistent surface water rights permit system which is different than the system just described for the Lower Rio Grande. The TCEQ administers both systems. River authorities, municipal utility districts, cities, irrigation districts, farmers, companies, and citizens hold about 8,000 permits statewide. The water is owned by the state, with rights granted to organizations and people to use prescribed amounts for prescribed purposes under prescribed conditions.

3.6.1. Historical Evolution Of Water Rights

The riparian doctrine was introduced in Texas during the rule of Spain and later Mexico and after independence in 1836, in a somewhat different form by the Republic of Texas. In 1840, the state of Texas adopted the common law of England with another variation of riparian water rights. The extent to which riparian rights allow large irrigation developments or other large amounts of use depends upon the laws in effect when the land was originally transferred from public to private ownership. Riparian rights are different depending on whether the land can be traced to land grants from Spain, Mexico, Republic of Texas, or the state of Texas.

The prior appropriation doctrine was adopted by legislative acts in 1889 and 1895. After 1895, public lands which transferred into private ownership no longer carried riparian water rights. Land already privately owned kept its riparian rights. At first, appropriation simply involved water users filing sworn statements with county clerks describing their use. Since 1913, more strictly administered procedures have been followed based on a statewide appropriation system administered by a centralized state agency.

All appropriation statutes recognized existing riparian rights. Riparian landowners could also acquire appropriative water rights. In 1926, the courts divided stream flow into *ordinary normal flow* and *flood flow*. Riparian rights were limited to normal flow and therefore are not applicable to flood waters impounded by reservoirs. Although difficult to apply in actual practice, this distinction was the basis for correlating riparian and appropriative rights from 1926 until the riparian rights were merged into the appropriative system pursuant to the Water Rights Adjudication Act of 1967.

Thus, an unmanageable system evolved, with various conflicting rights and many rights being unrecorded. The 1951-1957 drought motivated the previously noted Lower Rio Grande Valley Water Case which clearly demonstrated the impracticality of a purely judicial adjudication of water rights statewide. Thus, the Water Rights Adjudication Act was enacted by the Texas legislature in 1967, with the stated purpose being to require a recording of all claims for water rights, to limit the exercise of those claims to actual use, and to provide for the adjudication and administration of water rights. All riparian water rights were merged into the prior appropriation system, creating the present permit system applicable to all of Texas except the Lower Rio Grande. The water rights adjudication process required for transition to the permit system was initiated in 1968 and completed by the late 1980's.

3.6.2. Prior Appropriation Permit System

Water rights are granted by a state license, or permit, which allows the holder to divert a specified amount of water annually at a specific location, for a specific purpose, and to store water in reservoirs of specified capacity. Any organization or person may submit an application to the TCEQ for a new water right or to change an existing water right at any time. The TCEQ will approve the permit application if unappropriated water is available, a beneficial use of the water is contemplated, water conservation will be practiced, existing water rights are not impaired, and the water use is not detrimental to the public welfare. Proposed actions reflected in water right permit applications must be consistent with regional water plans.

The water authorized to be appropriated under the terms of the particular permit is not subject to further appropriation unless the permit is canceled. A permit may be canceled if water is not used during a 10-year period. Special term permits may also be issued allowing water use for specified periods of time. The Rio Grande and segments of other rivers are over-appropriated with no new rights for additional water use being granted.

However, unappropriated water is still available for appropriation in other regions of the state.

A permit holder has no actual title of ownership of the water but only a right to use the water. However, a water right can be sold, leased, or transferred to another person. The Lower Rio Grande Valley has been the only region of Texas with an active water market historically. In 1993, the legislature established a statewide water bank to be administered by the TWDB. Although transfers can be accomplished independently of the water bank, the program was created to encourage and facilitate water marketing, transfer, and reallocation.

The Texas Water Code lists beneficial uses in order of priority as follows: (1) domestic and municipal, (2) industrial, (3) irrigation, (4) mining, (5) hydroelectric, (6) navigation, (7) recreation and pleasure, and (8) other beneficial uses. These priorities are invoked only when a conflict exists between water use applications. Under the prior appropriation system, after a permit is in effect, priorities are based on the date specified in the permits. During the 1968-1980's adjudication process, priority dates were established based on historical water use. Since then, priorities are based on the dates that the permits are issued. In times of emergency, cities may take water even if non-municipal users are adversely affected, regardless of priority dates.

4. WATER RIGHTS ANALYSIS PACKAGE (WRAP) MODELING SYSTEM

The Water Rights Analysis Package (WRAP) computer modeling system is generalized for application to river/reservoir/use systems located anywhere in the world, with model-users developing input datasets for the particular river basin of concern. For applications in Texas, publicly available WRAP input datasets from the TCEQ Water Availability Modeling (WAM) System [28] are altered as appropriate to reflect proposed water management plans of interest, which could involve changes in water use or reservoir/river system operating practices, construction of new facilities, or other water management strategies.

WRAP simulates water resources development, management, regulation, and use in a river basin or multiple-basin region under a priority-based water allocation system. In WRAP terminology, a water right is a set of water use requirements, reservoir storage and conveyance facilities, hydropower projects, operating rules, and institutional arrangements for managing water resources. Stream flow and reservoir storage is allocated among users based on specified priorities, which can be defined in various ways. Simulation results are organized in optional formats including time sequences of many different variables, summaries, water budgets, frequency relationships, and various types of reliability indices.

The public domain WRAP software and documentation [29, 30, 31, 32, 33] may be downloaded free-of-charge from http://ceprofs.tamu. edu/ rwurbs/wrap.htm. The WRAP website links to TCEQ WAM and TWRI websites that provide reports, datasets, and other related information. Wurbs provides a concise summary of WRAP [34] and comparison with other generalized modeling systems [35].

WRAP modeling capabilities that have been routinely applied in the Texas WAM System consist of using a hydrologic period-of-analysis of about 60 years and monthly time step to perform water availability and reliability analyses for municipal, industrial, and agricultural water supply, environmental instream flow, hydroelectric power generation, and reservoir storage requirements. Recently developed additional WRAP modeling capabilities include: short-term conditional reliability modeling [36]; daily time step modeling capabilities that include flow forecasting and routing and disaggregation of monthly flows to daily; simulation of flood control reservoir system operations [31]; and salinity simulation [37]. Further improvements to WRAP currently underway, as of 2012, are focused on better integrating environmental flow requirements into comprehensive water management.

The generalized WRAP modeling system was developed and continues to be expanded at Texas A&M University (TAMU). Early versions dating back to the 1980's were developed under the sponsorship of the Texas Water Resources Institute (TWRI) and U.S. Department of Interior. Efforts at TAMU to expand and improve WRAP from 1996 through the present have been sponsored primarily by the TCEQ. However, the TWDB, U.S. Army Corps of Engineers, Brazos River Authority, and other agencies have funded specific improvements to WRAP. Model development has been an evolutionary process with extensive interactions between professionals from the agencies and consulting firms applying the model to specific river basins and university researchers responsible for improving the modeling methodology and computer software.

4.1. Texas Water Availability Modeling System

As previously noted, water management legislation known as Senate Bill 1 enacted by the Texas legislature in 1997 directed the water agencies to establish a regional planning process and a water availability modeling system. The Texas WAM System was implemented by the TCEQ, its partner agencies (TWDB and TPWD) and contractors (ten consulting engineering firms and two university research entities) during 1997-2002 pursuant to the 1997 Senate Bill 1. The WAM System has continuously to be improved and expanded since 2002 along with being routinely applied by the water management community [28].

The Water Availability Modeling (WAM) System consists of the generalized WRAP and input datasets for the 23 river basins of Texas. The WAM datasets include naturalized stream flows at a total of about 500 gauged sites, watershed parameters for distributing these flows to over 12,000 ungauged locations, 3,450 reservoirs, various other constructed infrastructure, operating plans that in many cases are quite complex, two international treaties, five interstate compacts, various contractual agreements, and water use requirements associated with about 8,000 water right permits reflecting two different water right systems.

Prior to creation of the WAM System, many water right permit holders incorrectly assumed that the amount of water specified in their permits would always be available to them. Senate Bill 1 required that the TCEQ notify all permit holders regarding the reliabilities associated with their permits. All water needs cannot be met during severe droughts.

The TCEQ requires that permit applicants, or their consultants, apply the WRAP/WAM system to assess supply reliabilities and flow and storage frequencies associated with their proposed actions and the impacts on all of the water users in the river basin. TCEQ staff applies the modeling system to evaluate the permit applications. The TWDB and regional planning groups apply the modeling system in their planning studies. The TCEQ requires that water right permit applications be consistent with regional plans. River authorities and other water suppliers apply the modeling system in operational planning studies.

4.2. Lessons Learned In Implementing The Wrap/Wam System

Developing and applying computer models have typically been viewed in terms of the engineering and scientific concepts and methods incorporated in the models. However, modeling has important institutional as well as technical dimensions. Lessons learned from development and application of the Texas WAM System demonstrate the importance of the following two institutional dimensions of river/reservoir system modeling.

1) Modeling water rights, contractual agreements, treaties, interstate compacts, and other complex institutional aspects of water resources development, management, allocation, and use may be a key consideration in developing and applying a modeling system.

2) Effective implementation of a modeling system may require a partnership effort of an entire water management community that includes water users, political officials, legislatures, environmental and other non-governmental special interest groups, government agencies, consulting firms, and university researchers.

The following general observations characterize the Texas experience in implementing a water availability modeling system.

- Droughts motivate political concern, improvements in water management, and development of computer modeling systems.

- Partnerships and consensus building are key aspects of water resources planning and management. Likewise, a water management community may work together to effectively implement a shared modeling system. Model development and application may be an institutional partnership effort.

- Administration of water allocation systems has become a central focus of river basin management. Regulatory and planning functions are integrally related. Shared modeling tools can facilitate integration of planning and regulatory functions.

- Modeling systems include computer programs, databases, organizations, people, and decision-making processes. Compilation and management of voluminous data is a central governing concern. A modeling system is constructed rather than just a model.

- Model development is a dynamic evolutionary process. As long as a computer simulation model such as WRAP continues to be applied, its development is never completed. Model development is a process of continual expanding and improving.

- Water availability modeling is essential for effective water management.

5. WATER ALLOCATION ISSUES

Allocating water resources that are highly variable both temporally and spatially among a myriad of water management entities and numerous water users within an institutional setting that has evolved historically over many years is necessarily complex. The following concerns highlighted by the Texas experience are illustrative of the numerous complexities in creating and administering water allocation systems.

For most of Texas, the water right permit system is administered without water master operations. Upon request, the TCEQ takes enforcement action to stop reported unauthorized water use in violation of water rights permits. However, water users are not closely monitored except during droughts or emergency conditions. This approach is similar to most western states. Several western states have water-master operations, but most states do not. The TCEQ during 2012-2013 is investigating the feasibility of expanding water master operations in Texas.

The TCEQ Lower Rio Grande Water Master Office maintains a precise accounting of water use, working closely with irrigators, cities, and the International Boundary and Water Commission. With completion of the adjudication process during the late 1980's, plans were developed to establish water-master operations in all of the major river basins of Texas. The South Texas Water-Master Program was created in the late 1980's with responsibilities for the Guadalupe, Nueces, and San Antonio River Basins. However, water users are reluctant to have requirements imposed upon them for installing meters and monitoring and regulating diversions. Political pressures have prevented the establishment of water-master offices in other river basins. However, the Texas Legislature in 2012 directed the TCEQ to solicit public input and develop recommendations for establishing water master operations for other river basins.

Since stream flow, evaporation, reservoir sedimentation, water use, and other factors are highly variable, and the future is unknown, water availability must be viewed from a reliability, likelihood, or percent-of-time perspective. Tradeoffs occur between the amount of water to commit for beneficial use and the level of reliability that can be achieved. Beneficial use of water is based on assuring a high level of reliability. However, if water commitments are limited as required to assure an extremely high level of reliability, the amount of stream flow available for beneficial use is constrained, and a greater proportion of the water flows to the ocean or is lost through reservoir evaporation. The optimal level of reliability varies with type of water use. Water allocation decisions necessarily require qualitative judgment in determining acceptable levels of reliability both in terms of the reliability of the proposed new or increased water use and the impacts on the reliabilities of all of the existing water users.

Many of the existing water rights adjudicated pursuant to the 1967 Water Rights Adjudication Act and the Lower Rio Grande Court Case as well as some recently established rights have supply reliabilities that are much lower than the water management community considers desirable. However, the TCEQ has applied more stringent criteria in approving newer water right permits or modifications to permits. In evaluating water right permit applications for agricultural irrigation, the TCEQ now applies the criterion that at least 75 percent of the proposed demand should be supplied at least 75 percent of the time as determined by the WRAP/WAM System. The TCEQ criterion for municipal permit applications is that 100 permit of the demand should be supplied 100 permit of the time based on the premises reflected in the WRAP/WAM model including historical hydrology. However, these criteria may be modified depending upon backup water supply sources. For example, with depleting groundwater reserves, a transition to a surface water source may be worthwhile, even if the reliability of the surface water source is low, if groundwater can still be used as a backup supply.

Although Texas and other western states are viewed as adopting the prior appropriation doctrine, strictly speaking a pure prior appropriation system is not feasible and does not exist. Although the effect may be very small, developing additional supplies for new users always affects downstream supplies. Also, in drought situations, water supply shortages are shared, to some degree, by water users regardless of the relative seniority of their rights. Sharing of water during drought typically depends on political negotiations, alternative demand management and supply augmentation measures available to different entities, and other factors in addition to the water rights permit system.

Assigning water right priorities to maintaining reservoir storage levels relative to diversion rights is another issue. Protecting reservoir inflows is crucial to providing a dependable water supply. Each drawdown could potentially be the beginning of a several-year drawdown that empties the reservoir. However, forcing appropriators, with rights junior to the rights of the reservoir owner, to curtail diversions to maintain inflows to an almost full, or even an almost empty, reservoir is difficult and often is not the optimal use of the water resource. If junior diversions are not curtailed, the reservoir will likely later refill anyway, without any supply shortages occurring. Handling of the storage aspect of water rights is not yet precisely defined in Texas except for the Lower Rio Grande. The Lower Rio Grande is simpler in this regard because essentially all of the water users are supplied by two large reservoirs. Elsewhere, numerous reservoirs are owned and operated by various entities in the same river basin.

Although some recently issued permits specify the amount of the diversion to be returned to the stream, most permits do not. Return flows can significantly impact the availability of water to downstream users. This issue is currently being addressed particularly as related to programs to encourage reuse of wastewater effluents.

From the perspective of hydrology and water resources management, groundwater and surface water are closely interrelated. Use of one resource often has significant impacts on the other. However, there is only limited governmental control over the use of groundwater in Texas. Consequently, conjunctive management of ground and surface water resources is difficult. Depleting groundwater reserves are forcing a shift toward greater groundwater regulation.

Water availability depends upon water quality as well as quantity. The water supply capabilities of several major river systems that include very large reservoirs in Texas and neighboring states are severely constrained by salinity originating from natural salt deposits in geologic formations underlying the upper watersheds of the rivers. Salinity simulation capabilities have been added to WRAP for assessing the impacts of salinity on water availability [37]. Salinity problems are addressed in the regional plans [24]. However, incorporation of salinity into the water allocation systems is a complex issue yet to be resolved.

The Texas Water Code has required consideration of environmental flow needs in the water rights permitting process since 1985. Such needs include maintenance of aquatic habitat and species, water quality, public recreation, wetlands, and freshwater inflows to bays and estuaries. Although such needs have been considered in issuing permits since 1985, most water rights in the state have been granted without specifying instream flow requirements. Under mandates enacted by the Texas Legislature in 2001 and 2007, developing methodologies for establishing environmental flow criteria, establishing requirements for each river reach in the state, and incorporating them into the water rights system is currently a major focus.

6. SUMMARY AND CONCLUSIONS

As water demands increase with population and economic growth, water allocation systems become increasing important worldwide. This chapter illustrates mechanisms for allocating water and highlights issues that must be addressed in their implementation. The Texas experience illustrates fundamental principles, issues, strategies, and complexities involved in developing and administering water allocation systems. Allocation of water resources among nations, states, regions, types of use, and numerous water users is a governing concern in water management in Texas as well as throughout the world.

Water allocation practices in Texas have evolved historically over centuries, with significant improvements occurring in recent years that continue to be refined. Texas shares water with Mexico and several neighboring states in the United States. Thousands of government agencies, cities, private companies, and citizens within Texas hold rights to use the waters of the state. Legal rights to use surface water differ from those for ground water. Surface water allocation for the Lower Rio Grande is different than for the rest of the state. With growing demands on limited water resources, expanding and refining water allocation systems has become a central governing concern in water management. The following observations regarding development and administration of water allocation systems in Texas are generally applicable in various other regions throughout the world.

- Water allocation systems include hierarchical systems of treaties and compacts between nations and states, ground water rights systems, surface water right systems, reservoir project ownership and operating practices, and contracts and agreements among a myriad of water management entities and water users. These water allocation mechanisms are overlapping and interconnected. Water allocation is integrally connected with other water resources planning and management activities.

- Water allocation practices evolve historically. In Texas, with growing demands on limited water resources, by the 1950's a disorganized myriad of water allocation practices had become a governing constraint to effective water management. Major change involving better water allocation mechanisms was necessary. The evolution of water allocation systems is continuing now and will continue in the future.

- Droughts motivate political action to improve water allocation systems. In Texas, major droughts led to milestone water management legislation being enacted in 1967 and 1997. The current drought that began in 2010 is expected to motivate additional legislative action that will contribute to the continued evolution of water allocation systems.

- The historical competition between agricultural, municipal, industrial, energy, and environmental water uses continues. There is a continuing shift from agricultural to municipal water use. Establishing and implementing environmental instream flow requirements is currently a major focus in Texas.

- Capabilities for assessing water availability and supply reliability within the framework of governing water allocation systems are essential for effective water resources planning and management. The Water Rights Analysis Package (WRAP) modeling system significantly contributes to effective water planning and allocation in Texas. WRAP is generalized for application to river/reservoir systems located anyplace in the world.

REFERENCES

1. Santos Roman DM. Systematization of water allocation systems: an engineering approach. PhD dissertation, Texas A&M University, College Station; 2005.

2. Rice L., White MD. Engineering Aspects of Water Law. Malabar: Krieger Publishing; 1991.

3. Wallace J., Wouters P., editors. Hydrology and Water Law: Bridging the Gap. London: International Water Association (IWA) Publishing; 2006.

4. Tofan C., Strambu S., editors. International Water Law, 2 volumes. Oisterwijk: ICA Press; 2008.

5. Beach LB., Mammer J., Hewitt JJ., Kaufman E., Kurki A., Oppenheimer JA., Wolf AT. Transboundary Freshwater Dispute Resolution: Theory, Practice, and Annotated References. New York: United Nations University Press; 2000.

6. Pricoli JD., Wolf AT. Managing and Transforming Water Conflicts. Cambridge : Cambridge University Press; 2009.

7. Wolf AT., Natharius JA., Danielson JJ., Ward BS., Pender JK. International River Basins of the World, International Journal of Water Resources Development 1999; 14(4) 387-427.

8. Grover VI., editor. Water: A Source of Conflict or Cooperation? Enfield: Science Publishers; 2007.

9. Barlow M. The Global Water Crisis and the Coming Battle for the Right to Water. Toronto: McClelland & Stewart, 2007.

10. Baillat A. International Trade in Water Rights: The Next Step. London: IWA Publishing; 2010.

11. Roth D., Boelens R., Zwarteveen M., editors. Liquid Relations: Contested Water Rights and Legal Complexities. New Brunswick: Rutgers University Press; 2005.

12. Selby J. Water, Power, and Politics in the Middle East. London: I.B. Tauris; 2003.

13. Getches DH. Water Law in a Nutshell. 4th Edition. St. Paul: West Publishing, 2009.

14. Beck RE., Kelley AK., editors. Waters and Water Rights, Cumulative Index, Volume 8, 2008 Edition. San Francisco: Matthew Bender (LexisNexis), 2008.

15. Draper, SE., editor. Model Water Sharing Agreements for the Twenty-First Century. Reston: American Society of Civil Engineers, 2002.

16. Eherat JW., editor. Riparian Water Regulations: Guidelines for Withdrawal Limitations and Permit Trading. Reston: American Society of Civil Engineers; 2002.

17. Dellapenna JW. Appropriative Rights Model Water Code, Reston: American Society of Civil Engineers; 2007.

18. Fischer D. The Law and Governance of Water Resources. Cheltenham: Edward Elgar Publishing; 2009.

19. Hu D. Water Rights: An International and Comparative Study. London: IWA Publishing; 2006.

20. Kissling-Nag I., Kuks S., editors. The Evolution of National Water Regimes in Europe: Transitions in Water Rights and Water Policies. London: Kluwer Publishers; 2004.

21. van Koppen B., Giordano M., Buttersworth J. Community-based Water Law and Water Resource Management Reform in Developing Countries. Oxfordshire: CAB International; 2007.

22. Longo PJ. Water Across Borders: Judicial Realities. In: Longo PJ., Yoskowitz DW. Eds. Water on the Great Plains: Issues and Policies. Lubbock: Texas Tech University Press; 2002.

23. Water Resources Planning: Manual of Water Supply Practices, Manual M50. Denver: American Water Works Association; 2001.

24. Water for Texas 2012. Austin: Texas Water Development Board; 2012.

25. Texas Instream Flow Studies: Technical Overview, Report 369. Austin: Texas Commission on Environmental Quality, Texas Parks and Wildlife Department, Texas Water Development Board; 2011.

26. Eaton DJ., Hurlbut D. Challenges in the Binational Management of Water Resources in the Rio Grande/Rio Bravo. U.S.-Mexican Policy Report No. 2. Austin: Lyndon B. Johnson School of Public Affairs, University of Texas; 1992.

27. Hall GE. High and Dry: the Texas-New Mexico Struggle for the Pecos River. Albuquerque: University of New Mexico Press; 2002.

28. Wurbs RA. Texas Water Availability Modeling System. Journal of Water Resources Planning and Management. American Society of Civil Engineers. 2005;131(4) 270-279.

29. Wurbs RA. Water Rights Analysis Package (WRAP) Modeling System Reference Manual. 9th Edition. TR-255. College Station: Texas Water Resources Institute; August 2012.

30. Wurbs RA. Water Rights Analysis Package (WRAP) Modeling System Users Manual. 9th Edition. TR-256. College Station: Texas Water Resources Institute; August 2012.

31. Wurbs RA., Hoffpauir RJ. Water Rights Analysis Package Daily Modeling System. TR-430, College Station: Texas Water Resources Institute; August 2012.

32. Wurbs RA. Fundamentals of Water Availability Modeling with WRAP. 6th Edition. TR-283, College Station: Texas Water Resources Institute; September 2011.

33. Wurbs RA. Salinity Simulation with WRAP. TR-317, College Station: Texas Water Resources Institute; July 2009.

34. Wurbs RA. Chapter 24 Water Rights Analysis Package (WRAP) Modeling System. In: Singh VP., Frevert D.K. (eds.) Watershed Models. Baton Raton: CRC Press; 2006.

35. Wurbs RA. Chapter 1 Generalized Models of River System Development and Management. In: Uhlig U. (ed.) Current Issues of Water Management. Rijeka: InTech 2011.

36. Wurbs RA., Schnier ST., Olmos HE. Short-Term Reservoir Storage Frequency Relationships. Journal of Water Resources Planning and Management. American Society of Civil Engineers. 2012;138(6).

37. Wurbs RA., Lee and CH. Salinity in Water Availability Modeling. Journal of Hydrology. Elsevier Science, 2011;407(2) 451-459.

Index